高等学校教材

材料力学性能

Mechanical Properties of Materials

刘春廷　马继　主编

Editor: Chunting Liu, Ji Ma

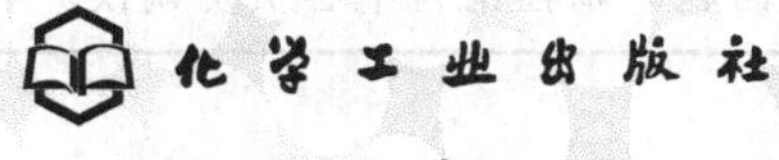

·北京·

本书是根据教育部最新颁布的课程教学基本要求和课程改革的精神编写的，以英文为表达形式，在内容和形式上有较大的更新，为材料性能学课程开展双语教学提供适用教材。

全书共十章，以工程材料的强度—硬度—塑性—韧性力学性能为主线，前七章详细阐述金属材料的力学性能，后三章分别阐述高分子材料、陶瓷材料和复合材料的力学性能，重点阐述工程材料在静载荷、冲击载荷和交变载荷及在环境介质（高温和腐蚀条件）作用下的力学性能，并从断裂力学的角度出发，重点阐述工程材料的抗断裂性能等。本书以阐述宏观规律为主，将宏观规律与微观机理相结合，同时强调理论与实际相联系。

本书作为机械类和材料类学生的专业基础课程材料性能学的教学用书，适用于48~64学时教学，主要面向机械类和材料类专本科学生，也可供近机类和近材料类专业选用，还可供有关工程技术人员学习参考。

图书在版编目 (CIP) 数据

材料力学性能 / 刘春廷，马继主编. —北京：化学工业出版社，2009. 9（2024. 9重印）
高等学校教材
ISBN 978-7-122-06462-2

Ⅰ. 材… Ⅱ. ①刘…②马… Ⅲ. 材料力学性质-高等学校-教材 Ⅳ. TB303

中国版本图书馆 CIP 数据核字（2009）第 140732 号

责任编辑：杨 菁 程树珍　　文字编辑：宋 薇
责任校对：陶燕华　　装帧设计：关 飞

出版发行：化学工业出版社（北京市东城区青年湖南街 13 号 邮政编码 100011）
印 装：北京科印技术咨询服务有限公司数码印刷分部
787mm×1092mm 1/16 印张 11½ 字数 299 千字 2024 年 9 月北京第 1 版第 2 次印刷

购书咨询：010-64518888　　售后服务：010-64518899
网 址：http://www.cip.com.cn
凡购买本书，如有缺损质量问题，本社销售中心负责调换。

定 价：29. 00 元

前　言

材料力学性能是高等院校机械类、材料类和近机械类、近材料类学生的一门重要的专业基础课。随着经济、科技和教育的国际化发展，中国走向世界的同时，世界也在向中国走来，而双语教学是今后中国教育改革的趋势和发展方向。本书在参考大量外文文献和外文教材的同时，依照中国人撰写教材和著作的逻辑思维以及分析问题和解决问题的方式方法，编写符合中国人自己思路的英文形式的教材，为双语教学的进一步实施提供物质基础。

全书共十章，以工程材料的强度—硬度—塑性—韧性力学性能为主线，前七章详细阐述金属材料的力学性能，后三章分别阐述高分子材料、陶瓷材料和复合材料的力学性能，阐述工程材料在静载荷、冲击载荷和交变载荷及在环境介质（高温和腐蚀条件）作用下的力学性能，并从断裂力学的角度出发，重点阐述工程材料的抗断裂性能等。本书以阐述宏观规律为主，将宏观规律与微观机理相结合，同时强调理论与实际相联系。

为配合学习，各章前提出学习目的，各章末附有本章小结、重要词汇、参考文献和思考题与习题，便于读者深入学习研究。本书力求语言简洁，信息量大，科学性、实用性强，内容新颖，引入新成果和新进展，有利于培养学生的创新意识，拓宽读者专业知识面，便于读者了解当前国内外材料力学性能研究动态和发展趋势。

在本书的编写过程中，美国加州大学（University of California, San Diego）Marc Andr′e Meyers 教授在提供原版英文教材等方面给予了很大的帮助，在此谨表深切的谢意！同时中国科学院金属研究所的胡壮麒院士、管恒荣研究员和孙晓峰研究员对本书提出了许多宝贵的意见, 在此表示由衷的感谢！

同时本书在编写过程中参考了已出版的各种文献和教材（见各章末的参考文献），并注意吸收各院校、研究所和企业的教学改革经验及科研成果，对此，谨向上述涉及的单位和个人表示衷心的感谢。

由于编者水平有限，加之时间仓促，书中若有不足之处，恳请广大读者和师生批评指正。

编　者
2009 年 6 月

Preface

Courses in the mechanical properties of materials are standard in both mechanical engineering and materials science/engineering curricula. These courses are taught, usually, at the junior or senior level. This book provides an introductory treatment of the mechanical properties of materials with a balanced mechanics—materials approach, which makes it suitable for both mechanical and materials engineering students.

The book covers mechanical properties of metals, polymers, ceramics, and composites and contains more than sufficient information for a one-semester course. It therefore enables the instructor to choose the path most appropriate to the class level (junior- or senior-level undergraduate) and background (mechanical or materials engineering). The book is organized into 10 chapters. Chapter 1 contains introductory information on materials that students with a previous course in the properties of materials should be familiar with. In addition, it enables those students unfamiliar with materials to "get up to speed." Chapter 2 through 7, on mechanical properties of metals, contains strength, hardness, toughness and ductility. Chapters 3 and 4, respectively, deal with fracture from a microstructural viewpoint (microfracure) and a macroscopic (primarily mechanical), and a broad treatment of plastic deformation and flow and fracture criteria are presented in Chapter 4. A detailed treatment of the fundamental mechanisms responsible for fatigue and creep, respectively, is presented in chapters 5 and 6. This is supplemented by a description of the principal testing and data analysis methods for these two phenomena. Chapters 7, on stress corrosion cracking (SCC) and hydrogen damage in metals, is essential to the understanding of environmentally assisted fracture in materials. Ceramics, polymers, composites, and intermetallics are nowadays important structural materials for advanced applications and are comprehensively covered in this book. In Chapters 8, 9 and 10, mechanical properties of nonmetals, including ceramics, polymers and composite materials are presented, respectively.

The objective is to include features in the book that will expedite the learning process. These learning aids include:

- Learning objectives;
- Numerous illustrations, now presented in the book, and photographs to help visualize what is being presented;
- End-of-chapter summary;
- Key terms and descriptions of key equations;
- Each chapter contains, at the end, a list of suggested reference; readers should consult these sources if they need to expand a specific point or if they want to broaden their knowledge in an area.
- All chapters contain solved examples and extensive lists of homework problems.These should be valuable tools in helping the student to grasp the concepts presented.

Full acknowledgment is here to all sources of tables and illustrations. We might have inadvertently forgotten to cite some of the sources in the final text; we sincerely apologize if we have failed to do so. The patient and competent typing of the manuscript by Jennifer Natelli, drafting by Jessica Mckinnis, and editorial help with text and problems by H.C. (Bryan) Chen and Elizabeth Kristofetz are gratefully acknowledged. we would like to acknowledge research support, over the years, from the U.S. Office of Naval Research, Oak Ridge National Laboratory, Los Alamos National Laboratory, and Sandia National Laboratories. we are also very thankful to his wife, Nivedita; son, Nikhilesh; and daughter, Kanika, for making it all worthwhile! Kanika's help in word processing is gratefully acknowledged. We acknowledge the continued support of the National Science Foundation (especially R.J. Reynik and B. Mac- Donald), the U.S. Army Research Office (especially G. Mayer, A. Crowson, K.Iyer, and E.Chen), and the Office of Naval Research. The inspiration provided by his grandfather, Jean-Pierre Meyers, and father, Henri Meyers, both metallurgists who devoted their lives to the profession, has inspired us. Professor Marc Andr'e Meyers from University of California, San Diego and Professor Krishan Kumar Chawla from University of Alabama at Birmingham generously supported the writing of the book. The help provided by Professor R.Skalak, director of the institute, is greatly appreciated. The Institute for Mechanics and Materials is supported by the National Science Foundation. The authors are grateful for the hospitality of Professor William D.Callister, Jr. from Department of Metallurgical Engineering, The University of Utah during the part of the preparation of the book.

Chunting Liu
Ji Ma
June, 2009

Contents

Chapter 1

Introduction

The successful utilization of materials requires that they satisfy a set of properties. These properties can be classified into thermal, optical, mechanical, physical, chemical, and nuclear, and they are intimately connected to the structure of materials. The structure, in its turn, is the result of synthesis and processing. A schematic framework that explains the complex relationships in the field of the mechanical behavior of materials, shown in Figure 1.1, is Thomas's iterative tetrahedron, which contains four principal elements: mechanical properties, characterization, theory, and processing. These elements are related, and changes in one are inseparably linked to changes in the others. For example, changes may be introduced by the synthesis and processing of, for instance, steel. The most common metal, steel has a wide range of strengths and ductilities (mechanical properties), which makes it the material of choice for numerous applications. While low carbon steel is used as reinforcing bars in concrete and in the body of automobiles, quenched and tempered high-carbon steel is used in more critical applications such as axles and gears. Cast iron, much more brittle, is used in a variety of applications, including automobile engine blocks. These different applications require, obviously, different mechanical properties of the material. The different properties of the three materials, resulting in differences in performance, are due to differences in the internal structure of the materials. The understanding of the structure comes from theory. The determination of the many aspects of the micro-, meso-, and macrostructure of materials is obtained by characterization. Low-carbon steel has a primarily ferritic structure, with some interspersed pearlite (a ferrite-cementite mixture). The high hardness of the quenched and tempered high-carbon steel is due to its martensitic structure (body-centered tetragonal). The relatively brittle cast iron has a structure resulting directly from solidification, without subsequent mechanical working such as hot rolling. How does one obtain low-carbon steel, quenched and tempered high-carbon steel, and cast iron? By different synthesis and processing routes. The low carbon steel is processed from the melt by a sequence of mechanical working operations. The high-carbon steel is synthesized with a greater concentration of carbon (>0.5%) than the low-carbon steel is (0.1%). Additionally, after mechanical processing, the high-carbon steel is rapidly cooled from a temperature of approximately 1000°C by throwing it into water or oil; it is then reheated to an intermediate temperature (tempering). The cast iron is synthesized with even higher carbon contents (about 2%). It is poured directly into the molds and allowed to solidify in them. Thus, no mechanical working, except for some minor machining, is needed. These interrelationships among structure, properties, and performance, and their modification by synthesis and processing, constitute the central theme of materials science and engineering. The tetrahedron of Figure 1.1 lists the principal processing methods, the most important theoretical approaches, and the most used characterization techniques in materials science today.

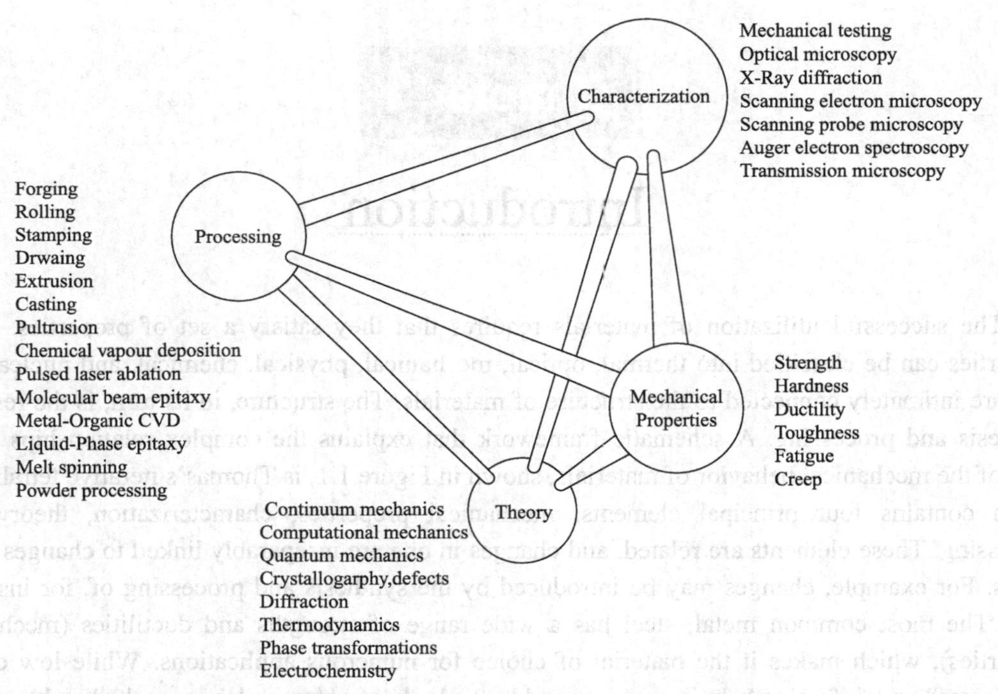

Figure 1.1 Iterative materials tetrahedron applied to mechanical behavior of materials. (After G.Thomas.)

The selection, processing, and utilization of materials have been part of human culture since its beginnings. Anthropologists refer to humans as "the toolmakers," and this is indeed a very realistic description of a key aspect of human beings responsible for their ascent and domination over other animals. It is the ability of humans to manufacture and use tools, and the ability to produce manufactured goods, that has allowed technological, cultural, and artistic progress and that has led to civilization and its development. Materials were as important to a Neolithic tribe in the year 10000 B.C. as they are to us today. The only difference is that today more complex synthetic materials are available in our society, while Neolithics had only natural materials at their disposal: wood, minerals, bones, hides, and fibers from plants and animals. Although these naturally occurring materials are still used today, they are vastly inferior in properties to synthetic materials.

Solid materials have been conveniently grouped into three basic classifications: metals, ceramics, and polymers. This scheme is based primarily on chemical makeup and atomic structure, and most materials fall into one distinct grouping or another, although there are some intermediates. In addition, there are the composites, combinations of two or more of the above three basic material classes. A brief explanation of these material types and representative properties is offered next.

Metals

Materials in this group are composed of one or more metallic elements (such as iron, aluminum, copper, titanium, gold, and nickel), and often also nonmetallic elements (for example, carbon, nitrogen, and oxygen) in relatively small amounts. Atoms in metals and their alloys are arranged in a very orderly manner (as discussed in Chapter 3), and in comparison to the ceramics and polymers, are relatively dense (Figure 1.2).With regard to mechanical characteristics, these materials are relatively stiff (Figure 1.3) and strong (Figure 1.4), yet are ductile (i.e., capable of large amounts of deformation without fracture), and are resistant to fracture (Figure 1.5), which accounts for their widespread use in structural applications. Metallic materials have large numbers of nonlocalized electrons; that is, these electrons are not bound to particular atoms. Many properties of metals are directly attributable to these electrons. For example, metals are extremely good conductors of electricity and heat, and are not transparent to visible light; a polished metal surface has a lustrous appearance. In addition, some of the metals (viz., Fe, Co, and Ni) have desirable magnetic properties.

In this book, the types and mechanical properties of metals and their alloys are discussed from Chapter 2 to Chapter 6.

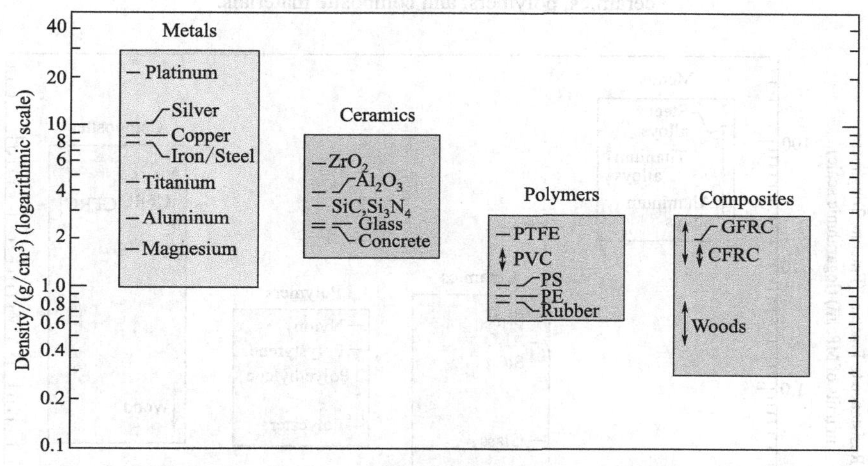

Figure 1.2 Bar-chart of room temperature density values for various metals, ceramics, polymers, and composite materials.

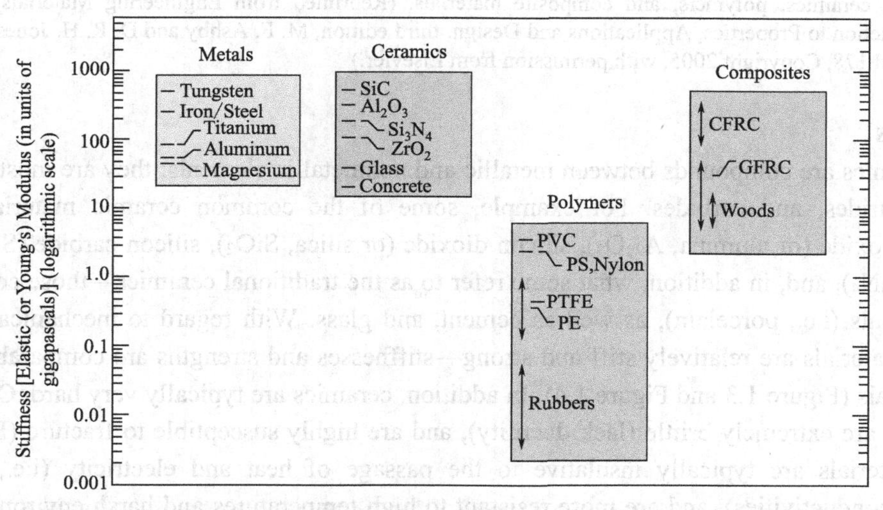

Figure 1.3 Bar-chart of room temperature stiffness (i.e., elastic modulus) values for various metals, ceramics, polymers, and composite materials.

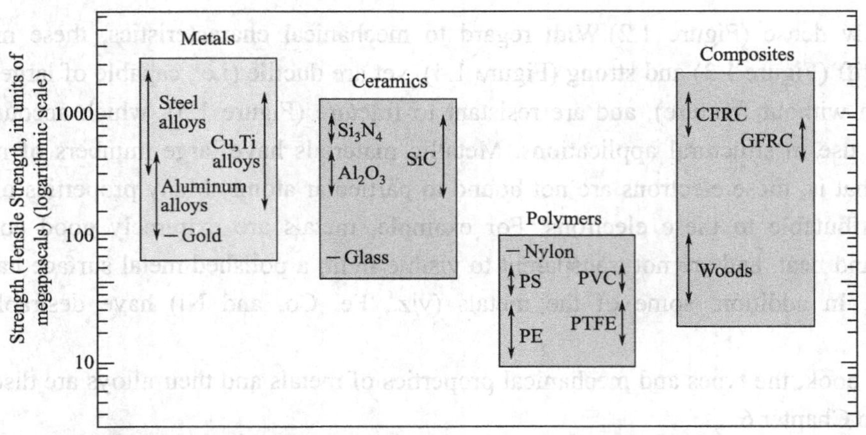

Figure 1.4 Bar-chart of room temperature strength (i.e., tensile strength) values for various metals, ceramics, polymers, and composite materials.

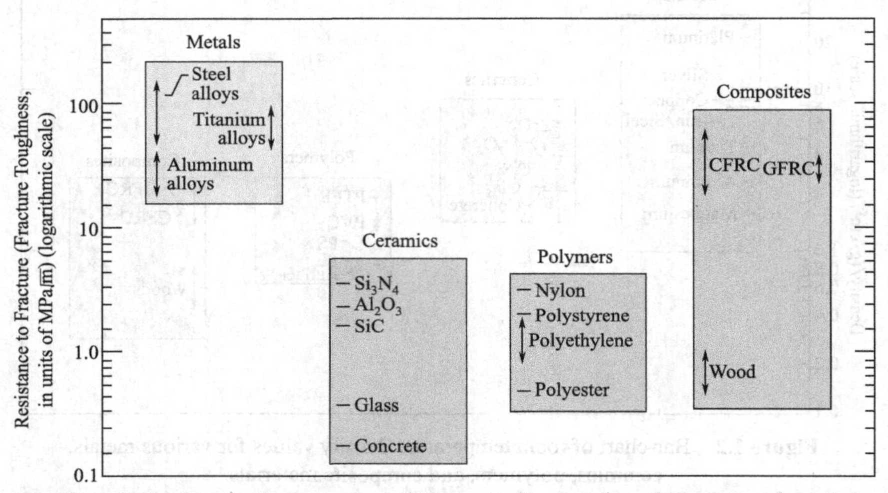

Figure 1.5 Bar-chart of room-temperature resistance to fracture (i.e., fracture toughness) for various metals, ceramics, polymers, and composite materials. (Reprinted from Engineering Materials 1: An Introduction to Properties, Applications and Design, third edition, M. F. Ashby and D. R. H. Jones, pages 177 and 178, Copyright 2005, with permission from Elsevier.)

Ceramics

Ceramics are compounds between metallic and nonmetallic elements; they are most frequently oxides, nitrides, and carbides. For example, some of the common ceramic materials include aluminum oxide (or alumina, Al_2O_3), silicon dioxide (or silica, SiO_2), silicon carbide (SiC), silicon nitride (Si_3N_4), and, in addition, what some refer to as the traditional ceramics—those composed of clay minerals (i.e., porcelain), as well as cement, and glass. With regard to mechanical behavior, ceramic materials are relatively stiff and strong—stiffnesses and strengths are comparable to those of the metals (Figure 1.3 and Figure 1.4). In addition, ceramics are typically very hard. On the other hand, they are extremely brittle (lack ductility), and are highly susceptible to fracture (Figure 1.5). These materials are typically insulative to the passage of heat and electricity (i.e., have low electrical conductivities), and are more resistant to high temperatures and harsh environments than metals and polymers. With regard to optical characteristics, ceramics may be transparent, translucent, or opaque (Figure 1.2), and some of the oxide ceramics (e.g., Fe_3O_4) exhibit magnetic behavior. Chapters 8 are devoted to discussions of the mechanical properties of these materials.

Polymers

Polymers include the familiar plastic and rubber materials. Many of them are organic compounds that are chemically based on carbon, hydrogen, and other nonmetallic elements (viz.O, N, and Si). Furthermore, they have very large molecular structures, often chain-like in nature that have a backbone of carbon atoms. Some of the common and familiar polymers are polyethylene (PE), nylon, poly(vinyl chloride) (PVC), polycarbonate (PC), polystyrene (PS), and silicone rubber. These materials typically have low densities (Figure 1.2), whereas their mechanical characteristics are generally dissimilar to the metallic and ceramic materials—they are not as stiff nor as strong as these other material types (Figure 1.3 and Figure 1.4). However, on the basis of their low densities, many times their stiffnesses and strengths on a per mass basis are comparable to the metals and ceramics. In addition, many of the polymers are extremely ductile and pliable (i.e., plastic), which means they are easily formed into complex shapes. In general, they are relatively inert chemically and unreactive in a large number of environments. One major drawback to the polymers is their tendency to soften and/or decompose at modest temperatures, which, in some instances, limits their use. Furthermore, they have low electrical conductivities and are nonmagnetic. Chapters 8 are devoted to discussions of the mechanical properties of polymeric materials.

Composites

A composite is composed of two (or more) individual materials, which come from the categories discussed above—viz., metals, ceramics, and polymers. The design goal of a composite is to achieve a combination of properties that is not displayed by any single material, and also to incorporate the best characteristics of each of the component materials. A large number of composite types exist that are represented by different combinations of metals, ceramics, and polymers. Furthermore, some naturally-occurring materials are also considered to be composites—for example, wood and bone. However, most of those we consider in our discussions are synthetic (or man-made) composites.

One of the most common and familiar composites is fiberglass, in which small glass fibers are embedded within a polymeric material (normally an epoxy or polyester). The glass fibers are relatively strong and stiff (but also brittle), whereas the polymer is ductile (but also weak and flexible). Thus, the resulting fiberglass is relatively stiff, strong, (Figure 1.3 and Figure 1.4) flexible, and ductile. In addition, it has a low density (Figure 1.2). Another of these technologically important materials is the "carbon fiber reinforced polymer" (or "CFRP") composite—carbon fibers that are embedded within a polymer. These materials are stiffer and stronger than the glass fiber-reinforced materials (Figure 1.3 and Figure 1.4), yet they are more expensive. The CFRP composites are used in some aircraft and aerospace applications, as well as high-tech sporting equipment (e.g., bicycles, golf clubs, tennis rackets, and skis/snowboards). Chapter 9 is devoted to a discussion of the mechanical properties of these interesting materials.

References

1. Ashby, M. F. and D. R. H. Jones, Engineering Materials 1, An Introduction to Their Properties and Applications, 3rd edition,

Butterworth-Heinemann,Woburn, UK, 2005

2. Ashby, M. F. and D. R. H. Jones, Engineering Materials 2, An Introduction to Microstructures, Processing and Design, 3rd edition, Butterworth-Heinemann,Woburn, UK, 2005
3. Askeland, D. R. and P. P. Phulé, The Science and Engineering of Materials, 5th edition, Nelson (a division of Thomson Canada), Toronto, 2006
4. Baillie, C. and L.Vanasupa, Navigating the Materials World, Academic Press, San Diego, CA, 2003
5. Flinn, R. A. and P. K. Trojan, Engineering Materials and Their Applications, 4th edition, John Wiley & Sons, New York, 1994
6. Jacobs, J. A. and T. F. Kilduff, Engineering Materials Technology, 5th edition, Prentice Hall PTR, Paramus, NJ, 2005
7. Mangonon, P. L., The Principles of Materials Selection for Engineering Design, Prentice Hall PTR, Paramus, NJ, 1999
8. McMahon, C. J., Jr., Structural Materials, Merion Books, Philadelphia, 2004
9. Murray, G. T., Introduction to Engineering Materials—Behavior, Properties, and Selection, Marcel Dekker, Inc., New York, 1993
10. Ralls, K. M., T. H. Courtney, and J. Wulff, Introduction to Materials Science and Engineering, John Wiley & Sons, New York, 1976
11. Schaffer, J. P., A. Saxena, S. D. Antolovich, T. H. Sanders, Jr., and S. B. Warner, The Science and Design of Engineering Materials, 2nd edition, WCB/McGraw-Hill, New York, 1999
12. Shackelford, J. F., Introduction to Materials Science for Engineers, 6th edition, Prentice Hall PTR, Paramus, NJ, 2005
13. Smith, W. F. and J. Hashemi, Principles of Materials Science and Engineering, 4th edition, McGraw-Hill Book Company, New York, 2006
14. Van Vlack, L. H., Elements of Materials Science and Engineering, 6th edition, Addison-Wesley Longman, Boston, MA, 1989
15. White, M. A., Properties of Materials, Oxford University Press, New York, 1999
16. J. F. Shackelford. Introduction to Materials Science for Engineers, 4th ed. Upper Saddle River, NJ: Prentice Hall, 1996
17. W. F. Smith. Principles of Materials Science and Engineering, 3rd ed. New York: McGraw Hill, 1996
18. D. R. Askeland and P. Phule. The Science and Engineering of Materials, 4th ed. Pacific Grove, CA: Thomson, 2003
19. W. D. Callister. Jr. Materials Science and Engineering, 4th ed. New York: Wiley, 2003

Chapter 2

Mechanical Properties of Metals

Learning Objectives:

After studying this chapter you should be able to do the following:

1. Define engineering stress and engineering strain.
2. State Hooke's law, and note the conditions under which it is valid.
3. Define Poisson's ratio.
4. Given an engineering stress-strain diagram, determine:
   a. the modulus of elasticity,
   b. the yield strength,
   c. the tensile strength,
   d. estimate the percent elongation.
5. For the tensile deformation of a ductile cylindrical specimen, describe changes in specimen profile to the point of fracture.
6. Compute ductility in terms of both percent elongation and percent reduction of area for a material that is loaded in tension to fracture.
7. Give brief definitions of and the units for modulus of resilience and toughness (static).
8. For a specimen being loaded in tension, given the applied load, the instantaneous cross-sectional dimensions, as well as original and instantaneous lengths, is able to compute true stress and true strain values.
9. Name the two most common hardness-testing techniques; note two differences between them.
10. a. Name and briefly describe the two different microindentation hardness testing techniques,
    b. cite situations for which these techniques are generally used.

2.1 Introduction

Many materials, when in service, are subjected to forces or loads; examples include the aluminum alloy from which an airplane wing is constructed and the steel in an automobile axle. In such situations it is necessary to know the characteristics of the material and to design the member from which it is made such that any resulting deformation will not be excessive and fracture will not occur. The mechanical behavior of a material reflects the relationship between its response and deformation to an applied load or force. Important mechanical properties are strength, hardness, ductility, and stiffness.

The mechanical properties of materials are ascertained by performing carefully designed laboratory experiments that replicate as nearly as possible the service conditions. Factors to be considered include the nature of the applied load and its duration, as well as the environmental conditions. It is possible for the load to be tensile, compressive, or shear, and its magnitude may be constant with time, or it may fluctuate continuously. Application time may be only a fraction of a second, or it may extend over a period of many years. Service temperature may be an important factor. Mechanical properties are of concern to a variety of parties (e.g., producers and consumers of materials, research organizations, and government agencies) that have differing interests. Consequently, it is imperative that there be some consistency in the manner in which tests are conducted, and in the interpretation of their results. This consistency is accomplished by using standardized testing techniques. Establishment and publication of these standards are often coordinated by professional societies. In the United States the most active organization is the American Society for Testing and Materials (ASTM).

The role of structural engineers is to determine stresses and stress distributions within members that are subjected to well-defined loads. This may be accomplished by experimental testing techniques or by theoretical and mathematical stress analysis. These topics are treated in traditional stress analysis and strength of materials texts. Materials are frequently chosen for structural applications because they have desirable combinations of mechanical characteristics. The present discussion is confined primarily to the mechanical behavior of metals; polymers and ceramics are treated separately because they are, to a large degree, mechanically dissimilar to metals. This chapter discusses the stress–strain behavior of metals and the related mechanical properties, and also examines other important mechanical characteristics. Discussions of the microscopic aspects of deformation mechanisms and methods to strengthen and regulate the mechanical behavior of metals are deferred to later chapters.

2.2 Concepts Of Stress And Strain

If a load is static or changes relatively slowly with time and is applied uniformly over a cross section or surface of a member, the mechanical behavior may be ascertained by a simple stress-strain test; these are most commonly conducted for metals at room temperature. There are three principal ways in which a load may be applied: namely, tension, compression, and shear [Figures 2.1(a), (b), (c)]. In engineering practice many loads are torsional rather than pure shear; this type of loading is illustrated in Figure 2.1(d). One of the most common mechanical stress–strain tests is performed in *tension*. As will be seen, the tension test can be used to ascertain several mechanical properties of materials that are important in design. A specimen is deformed, usually to fracture, with a gradually increasing tensile load that is applied uniaxially along the long axis of a specimen. Normally, the cross section is circular, but rectangular specimens are also used. This "dogbone" specimen configuration was chosen so that, during testing, deformation is confined to the narrow center region (which has a uniform cross section along its length), and, also, to reduce the likelihood of fracture at the ends of the specimen.

Figure 2.2 shows a cylindrical specimen being stressed in a machine that tests materials for tensile strength. The upper part of the specimen is screwed to the crosshead of the machine. The coupled rotation of the two lateral screws causes the crosshead to move. The load cell is a transducer that measures the load and sends it to a recorder; the increase in length of the specimen can be read by strain gages, extensometers, or, indirectly, from the velocity of motion of the crosshead.

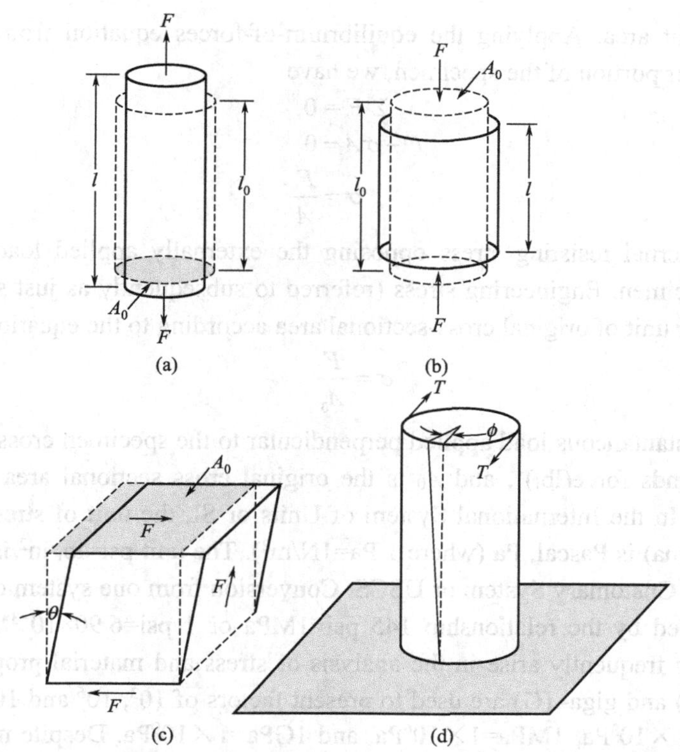

Figure 2.1 (a) Schematic illustration of how a tensile load produces an elongation and positive linear strain. Dashed lines represent the shape before deformation; solid lines, after deformation. (b) Schematic illustration of how a compressive load produces contraction and a negative linear strain. (c) Schematic representation of shear strain, where $\gamma = \tan\theta$. (d) Schematic representation of torsion deformation (i.e., angle of twist) produced by an applied torque T.

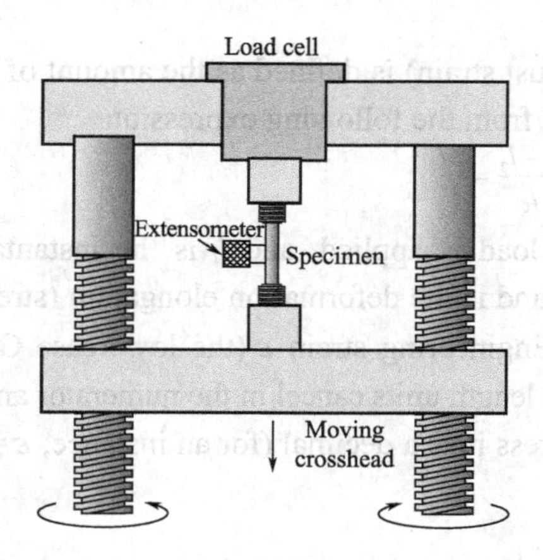

Figure 2.2 Schematic representation of the apparatus used to conduct tensile stress-strain tests. The specimen is elongated by the moving crosshead; load cell and extensometer measure, respectively, the magnitude of the applied load and the elongation.

Another type of machine, called a servohydraulic machine, is also used. Assuming that at a certain moment the force applied on the specimen by the machine is F, there will be a tendency to "stretch" the specimen, breaking the internal bonds. This breaking tendency is opposed by internal reactions, called *stresses*. The best way of visualizing stresses is by means of the method of analysis used in the mechanics of materials: The specimen is "sectioned" and the missing part is replaced by the forces that it exerts on the other parts. This procedure is indicated in the figure, and the "resistance" is uniformly distributed over the normal section. The normal stress σ is defined as this "resistance" per unit area. Applying the equilibrium-of-forces equation from the mechanics of materials to the lower portion of the specimen, we have

$$\Sigma F = 0$$

$$F - \sigma A = 0$$

$$\sigma = \frac{F}{A} \tag{2.1}$$

This is the internal resisting stress opposing the externally applied load and avoiding the breaking of the specimen. Engineering stress (referred to subsequently as just stress) is defined as the applied force per unit of original cross-sectional area according to the equation

$$\sigma = \frac{F}{A_0} \tag{2.2}$$

in which F is the instantaneous load applied perpendicular to the specimen cross section, in units of Newton (N) or pounds force(lb_f) , and A_0 is the original cross sectional area before any load is applied (m^2 or in^2). In the International System of Units or SI, the unit of stress σ (the lowercase Greek character sigma) is Pascal, Pa (where 1 Pa=1N/m^2). The unit psi=lb_f/in^2 is used for the stress in the United States Customary System or USCS. Conversion from one system of stress units to the other is accomplished by the relationship 145 psi=1MPa or 1 psi=6.90×10^{-3}MPa. Because large numerical quantities frequently arise in the analysis of stress and material properties, the prefixes kilo- (k), mega- (M) and giga- (G) are used to present factors of 10^3, 10^6 and 10^9, respectively. For an instance, 1kPa=1×10^3Pa, 1MPa=1×10^6Pa, and 1GPa=1×10^9Pa. Despite mixing formats with the SI, it is also conventional to use the prefixes kilo- and mega- when representing large number in the USCS. Mechanical engineers abbreviate the stresses 1×10^3 psi as 1ksi (without the "p"), and 1×10^6 psi as 1psi (with the "p"). In the USCS, the unit of 1 billion psi (Gpsi) is unrealistically large for calculations involving materials and stresses in mechanical engineering, and so it is not conventionally used.

Engineering strain (subsequently called just strain) is defined as the amount of stretching that occurs per unit of original length and calculated from the following expression

$$\varepsilon = \frac{l_i - l_0}{l_0} = \frac{\Delta l}{l_0} \tag{2.3}$$

in which l_0 is the original length before any load is applied, and l_i is the instantaneous length. Sometimes the quantity $l_i - l_0$ is denoted as Δl , and is the deformation elongation (stretch) or change in length at some instant when F is applied. Engineering strain ε (the lowercase Greek character epsilon) is a dimensionless quantity, because the length units cancel in the numerator and denominator. Strain is generally very small, and you can express it as a decimal (for an instance, ε=0.005) or as a percent (ε=0.5%).

Compression Tests

Compression stress-strain tests may be conducted in a manner similar to the tensile test, except that the force is compressive and the specimen contracts along the direction of the stress. Equation 2.2 and Equation 2.3 are utilized to compute compressive stress and strain, respectively. By convention, a compressive force is taken to be negative, which yields a negative stress. Furthermore, since l_0 is greater than l_i, compressive strains computed from Equation 2.3 are necessarily also negative. Tensile tests are more common because they are easier to perform; also, for most materials used in structural applications, very little additional information is obtained from compressive tests. Compressive tests are used when a material's behavior under large and permanent (i.e., plastic) strains is desired, as in manufacturing applications, or when the material is brittle in tension.

Shear and Torsion Tests

For tests performed using a pure shear force as shown in Figure 2.1(c), the shear stress τ is computed according to

$$\tau = \frac{F}{A_0} \tag{2.4}$$

where F is the load or force imposed parallel to the upper and lower faces, each of which has an area of A_0. The shear strain γ is defined as the tangent of the strain angle θ, as indicated in the figure 2.1(c).The units for shear stress and strain are the same as for their tensile counterparts.

Torsion is a variation of pure shear, wherein a structural member is twisted in the manner of Figure 2.1(d); torsion forces produce a rotational motion about the longitudinal axis of one end of the member relative to the other end. Examples of torsion are found for machine axles and drive shafts, and also for twist drills. Torsion tests are normally performed on cylindrical solid shafts or tubes. A shear stress τ is a function of the applied torque T, whereas shear strain γ is related to the angle of twist, ϕ in Figure 2.1(d).

2.3 Stress-Strain Behavior

Figure 2.3 shows two types of engineering-stress-strain curves with uniaxial tensile stress. The first exhibits a yield point, while the second does not. Many parameters are used to describe the various features of these curves. First, there is the elastic limit. A proportional limit is also sometimes defined (e); it corresponds to the stress at which the curve deviates from linearity. There is a drop in yield, an upper (A) and a lower (C) yield point defined in Fig. 2.3 (b). The lower yield point depends on the machine stiffness. In some case, since it is difficult to determine the maximum stress for which there is no permanent deformation, the 0.2% offset yield stress (point C in the figure) is commonly used instead; it corresponds to a permanent strain of 0.2% after unloading. The maximum engineering stress is called the ultimate tensile stress (σ_b); it corresponds to point B in

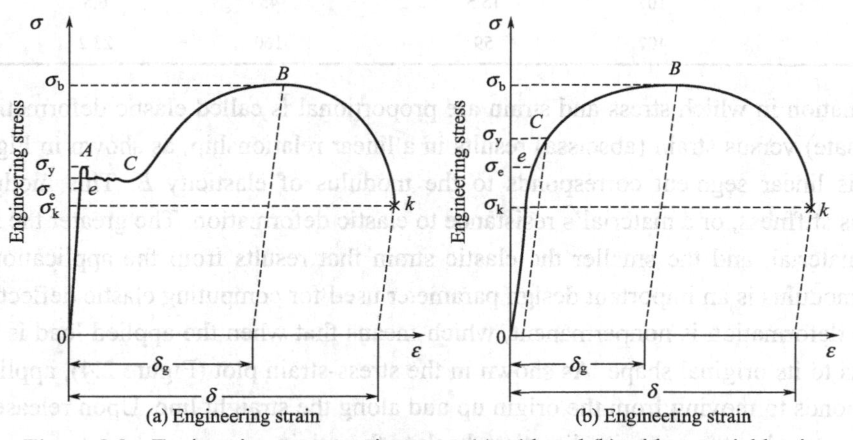

Figure 2.3 Engineering-stress-strain curves (a) with and (b) without a yield point

Figure 2.3. Beyond the ultimate tensile stress (σ_b), the engineering stress drops until the rupture stress (σ_k) is reached. The uniform strain corresponds to the plastic strain that takes place uniformly in the specimen. Beyond that point (*B*), necking occurs. Necking is treated in detail in Section 2.7. *k* is the strain-to-failure. Additional parameters can be obtained from the stress-strain curve:

(1) The elastic energy absorbed by the specimen (the area under the elastic portion of the curve) is called resilience;

(2) the total energy absorbed by the specimen during deformation, up to fracture (the area under the whole curve), is called work of fracture. The strain rate undergone by the specimen, $\sigma_e = d\varepsilon_e/dt$, is equal to the crosshead velocity, divided by the initial length l_0 of the specimen.

Elastic Deformation

The degree to which a structure deforms or strains depends on the magnitude of an imposed stress. For most metals that are stressed in tension and at relatively low levels, stress and strain are proportional to each other through the relationship

$$\sigma = E\varepsilon \tag{2.5}$$

This is known as Hooke's law—relationship between engineering stress and engineering strain for elastic deformation (tension and compression), and the constant of proportionality *E* (GPa or psi) is the modulus of elasticity, or Young's modulus. For most typical metals the magnitude of this modulus ranges between 45GPa (6.5Mpsi), for magnesium, and 407GPa (59Mpsi), for tungsten. Modulus of elasticity values for several metals at room temperature are presented in Table 2.1.

Table 2.1 Room-Temperature Elastic and Shear Moduli, and Poisson's Ratio for Various Metal Alloys

Metal Alloy	Modulus of Elasticity		Shear Modulus		Poisson's Ratio
	GPa	1×10^6 psi	GPa	1×10^6 psi	
Aluminum	69	10	25	3.6	0.33
Brass	97	14	37	5.4	0.34
Copper	110	16	46	6.7	0.34
Magnesium	45	6.5	17	2.5	0.29
Nickel	207	30	76	11.0	0.31
Steel	207	30	83	12.0	0.30
Titanium	107	15.5	45	6.5	0.34
Tungsten	407	59	160	23.2	0.28

Deformation in which stress and strain are proportional is called elastic deformation; a plot of stress (ordinate) versus strain (abscissa) results in a linear relationship, as shown in Figure 2.5. The slope of this linear segment corresponds to the modulus of elasticity *E*. This modulus may be thought of as stiffness, or a material's resistance to elastic deformation. The greater the modulus, the stiffer the material, and the smaller the elastic strain that results from the application of a given stress. The modulus is an important design parameter used for computing elastic deflections.

Elastic deformation is nonpermanent, which means that when the applied load is released, the piece returns to its original shape. As shown in the stress-strain plot (Figure 2.4), application of the load corresponds to moving from the origin up and along the straight line. Upon release of the load, the line is traversed in the opposite direction, back to the origin.

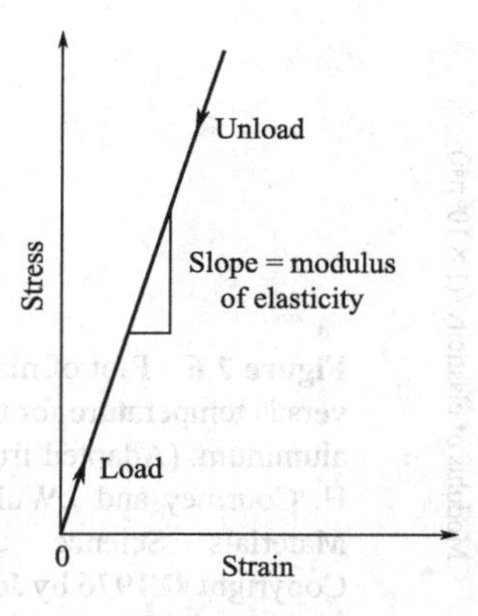

Figure 2.4 Schematic stress-strain diagrams showing linear elastic deformation for loading and unloading cycles.

On an atomic scale, macroscopic elastic strain is manifested as small changes in the interatomic spacing and the stretching of interatomic bonds. As a consequence, the magnitude of the modulus of elasticity is a measure of the resistance to separation of adjacent atoms, that is, the interatomic bonding forces. Furthermore, this modulus is proportional to the slope of the interatomic force-separation curve at the equilibrium spacing:

$$E \propto \left(\frac{dF}{dr}\right)_{r_0} \tag{2.6}$$

Figure 2.5 shows the force-separation curves for materials having both strong and weak interatomic bonds; the slope at r_0 is indicated for each.

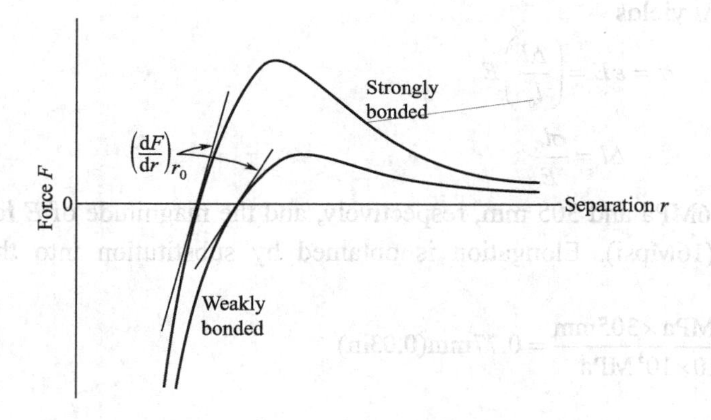

Figure 2.5 Force versus interatomic separation for weakly and strongly bonded atoms. The magnitude of the modulus of elasticity is proportional to the slope of each curve at the equilibrium interatomic separation r_0.

Values of the modulus of elasticity for ceramic materials are about the same as for metals; for polymers they are lower. These differences are a direct consequence of the different types of atomic bonding in the three materials types. Furthermore, with increasing temperature, the modulus of elasticity diminishes, as is shown for several metals in Figure 2.6.

As would be expected, the imposition of compressive, shear, or torsion stresses also evokes elastic behavior. The stress-strain characteristics at low stress levels are virtually the same for both tensile and compressive situations, to include the magnitude of the modulus of elasticity. Shear stress and strain are proportional to each other through the expression

$$\tau = G\gamma \tag{2.7}$$

where G is the shear modulus, the slope of the linear elastic region of the shear stress–strain curve. Table 2.1 also gives the shear moduli for a number of the common metals.

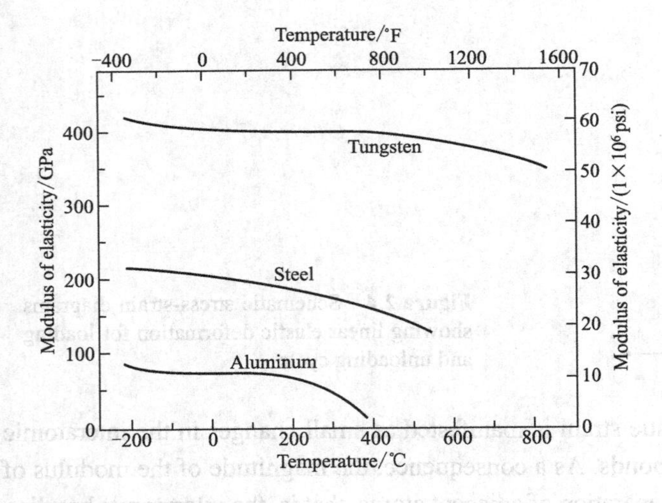

Figure 2.6 Plot of modulus of elasticity versus temperature for tungsten, steel, and aluminum. (Adapted from K. M. Ralls, T. H. Courtney and J.Wulff, Introduction to Materials Science and Engineering. Copyright © 1976 by John Wiley & Sons, New York. Reprinted by permission of John Wiley & Sons, Inc.)

Example Problem 2.1

Elongation (Elastic) **Computation**

A piece of copper originally 305 mm (12 in) long is pulled in tension with a stress of 276MPa (40000 psi). If the deformation is entirely elastic, what will be the resultant elongation?

Solution

Since the deformation is elastic, strain is dependent on stress according to Equation 2.5. Furthermore, the elongation Δl is related to the original length through Equation 2.2. Combining these two expressions and solving for Δl yields

$$\sigma = \varepsilon E = \left(\frac{\Delta l}{l_0}\right) E$$

$$\Delta l = \frac{\sigma l_0}{E}$$

The values of σ and l_0 are given as 276MPa and 305 mm, respectively, and the magnitude of E for copper from Table 2.1 is 110GPa (16Mpsi). Elongation is obtained by substitution into the expression above as

$$\Delta l = \frac{276\text{MPa} \times 305\text{mm}}{110 \times 10^3\ \text{MPa}} = 0.77\text{mm}(0.03\text{in})$$

2.4 Anelasticity (or Viscoelasticity)

Up to this point, it has been assumed that elastic deformation is time independent—that is, that an applied stress produces an instantaneous elastic strain that remains constant over the period of time the stress is maintained. It has also been assumed that upon release of the load the strain is totally recovered—that is, that the strain immediately returns to zero. In most engineering materials, however, there will also exist a time-dependent elastic strain component. That is, elastic deformation will continue after the stress application, and upon load release some finite time is required for complete recovery. This time-dependent elastic behavior is known as anelasticity, and it is due to time-dependent microscopic and atomistic processes that are attendant to the deformation. For metals the anelastic component is normally small and is often neglected.

However, for some polymeric materials its magnitude is significant; in this case it is termed viscoelastic behavior, which is the discussion topic of Section 9.4.

2.5 Elastic Properties of Materials

When a tensile stress is imposed on a metal specimen, an elastic elongation and accompanying strain ε_z result in the direction of the applied stress (arbitrarily taken to be the z direction), as indicated in Figure 2.7. As a result of this elongation, there will be constrictions in the lateral (x and y) directions perpendicular to the applied stress; from these contractions, the compressive strains ε_x and ε_y may be determined. If the applied stress is uniaxial (only in the z direction), and the material is isotropic, then $\varepsilon_x = \varepsilon_y$. A parameter termed Poisson's ratio ν is defined as the ratio of the lateral and axial strains, or

$$\nu = -\frac{\varepsilon_x}{\varepsilon_z} = -\frac{\varepsilon_y}{\varepsilon_z} \tag{2.8}$$

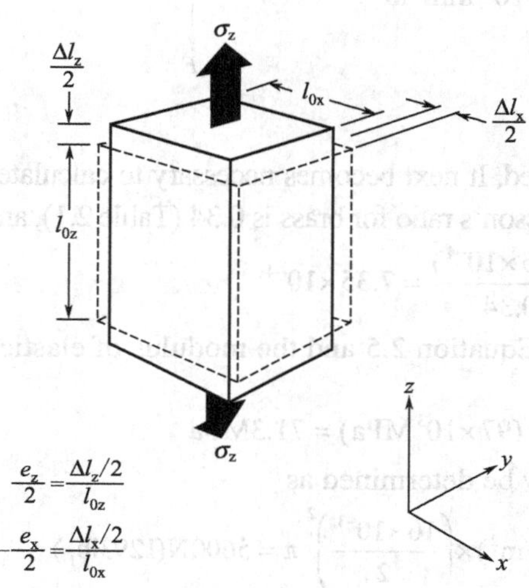

Figure 2.7 Axial (z) elongation (positive strain) and lateral (x and y) contractions (negative strains) in response to an imposed tensile stress. Solid lines represent dimensions after stress application; dashed lines, before.

The negative sign is included in the expression so that ν will always be positive, since ε_x and ε_z will always be of opposite sign. Theoretically, Poisson's ratio for isotropic materials should be; furthermore, the maximum value for (or that value for which there is no net volume change) is 0.50. For many metals and other alloys, values of Poisson's ratio range between 0.25 and 0.35. Table 2.1 shows ν values for several common metallic materials.

For isotropic materials, shear and elastic moduli are related to each other and to Poisson's ratio according to

$$E = 2G(1+\nu) \tag{2.9}$$

In most metals G is about $0.4E$; thus, if the value of one modulus is known, the other may be approximated.

Many materials are elastically anisotropic; that is, the elastic behavior (e.g., the magnitude of E) varies with crystallographic direction. For these materials the elastic properties are completely characterized only by the specification of several elastic constants, their number depending on characteristics of the crystal structure. Even for isotropic materials, for complete characterization of the elastic properties, at least two constants must be given. Since the grain orientation is random in most polycrystalline materials, these may be considered to be isotropic; inorganic ceramic glasses are also isotropic. The remaining discussion of mechanical behavior assumes isotropy and polycrystalline because such is the character of most engineering materials.

Example Problem 2.2

Computation of Load to Produce Specified Diameter Change

A tensile stress is to be applied along the long axis of a cylindrical brass rod that has a diameter of 10mm (0.4 in). Determine the magnitude of the load required to produce a 2.5×10^{-3}mm (10^{-4}in) change in diameter if the deformation is entirely elastic.

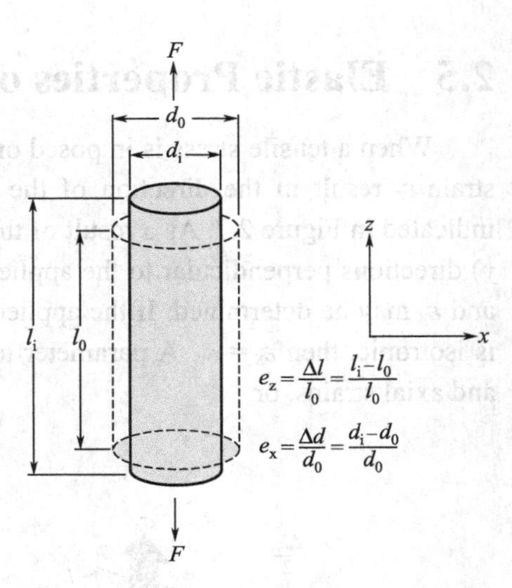

Solution

This deformation situation is represented in the accompanying drawing. When the force F is applied, the specimen will elongate in the z direction and at the same time experience a reduction in diameter, Δd, of 2.5×10^{-3}mm in the x direction. For the strain in the x direction

$$\varepsilon_x=\frac{\Delta d}{d_0}=\frac{-2.5\times10^{-3}\,\text{mm}}{10\text{mm}}=-2.5\times10^{-4}$$

Which is negative, since the diameter is reduced. It next becomes necessary to calculate the strain in the z direction using Equation 2.8. The value for Poisson's ratio for brass is 0.34 (Table 2.1), and thus

$$\varepsilon_z=-\frac{\varepsilon_x}{\nu}=-\frac{(-2.5\times10^{-4})}{0.34}=7.35\times10^{-4}$$

The applied stress may now be computed using Equation 2.5 and the modulus of elasticity, given in Table 2.1 as 97GPa (14Mpsi), as

$$\sigma=\varepsilon_z E=(7.35\times10^{-4})\times(97\times10^3\,\text{MPa})=71.3\text{MPa}$$

Finally, from Equation 2.1, the applied force may be determined as

$$F=\sigma A_0=\sigma\left(\frac{d_0}{2}\right)^2\pi=(71.3\times10^6\,\text{N/m}^2)\times\left(\frac{10\times10^{-3}}{2}\right)^2\pi=5600\text{N}(1293\text{lb}_\text{f})$$

Plastic Deformation

For most metallic materials, elastic deformation persists only to strains of about 0.005. As the material is deformed beyond this point, the stress is no longer proportional to strain (Hooke's law, Equation 2.5, ceases to be valid), and permanent, nonrecoverable, or plastic deformation occurs. Figure 2.8(a) plots schematically the tensile stress-strain behavior into the plastic region for a typical metal. The transition from elastic to plastic is a gradual one for most metals; some curvature results at the onset of plastic deformation, which increases more rapidly with rising stress.

From an atomic perspective, plastic deformation corresponds to the breaking of bonds with original atom neighbors and then reforming bonds with new neighbors as large numbers of atoms or molecules move relative to one another; upon removal of the stress they do not return to their original positions. The mechanism of this deformation is different for crystalline and amorphous materials. For crystalline solids, deformation is accomplished by means of a process called slip, which involves the motion of dislocations. Plastic deformation in noncrystalline solids (as well as liquids) occurs by a viscous flow mechanism.

2.6 Tensile Properties

2.6.1 Yield Strength

Most structures are designed to ensure that only elastic deformation will result when a stress is applied. A structure or component that has plastically deformed, or experienced a permanent change in shape, may not be capable of functioning as intended. It is therefore desirable to know the stress level at which plastic deformation begins, or where the phenomenon of yielding occurs. For metals that experience this gradual elastic-plastic transition, the point of yielding may be determined as the initial departure from linearity of the stress-strain curve; this is sometimes called the proportional limit, as indicated by point P in Figure 2.8 (a). In such cases the position of this point may not be determined precisely. As a consequence, a convention has been established wherein a straight line is constructed parallel to the elastic portion of the stress-strain curve at some specified strain offset, usually 0.002. The stress corresponding to the intersection of this line and the stress-strain curve as it bends over in the plastic region is defined as the yield strength σ_y. This is demonstrated in Figure 2.8(a). Of course, the units of yield strength are MPa or psi. For those materials having a nonlinear elastic region (Figure 2.6), use of the strain offset method is not possible, and the usual practice is to define the yield strength as the stress required to produce some amount of strain (e.g., $\varepsilon = 0.005$). Some steels and other materials exhibit the tensile stress-strain behavior as shown in Figure 2.8(b). The elastic-plastic transition is very well defined and occurs abruptly in what is termed a yield point phenomenon. At the upper yield point, plastic deformation is initiated with an actual decrease in stress. Continued deformation fluctuates slightly about some constant stress value, termed the lower yield point; stress subsequently rises with increasing strain. For metals that display this effect, the yield strength is taken as the average stress that is associated with the lower yield point, since it is well defined and relatively insensitive to the testing procedure. Thus, it is not necessary to employ the strain offset method for these materials.

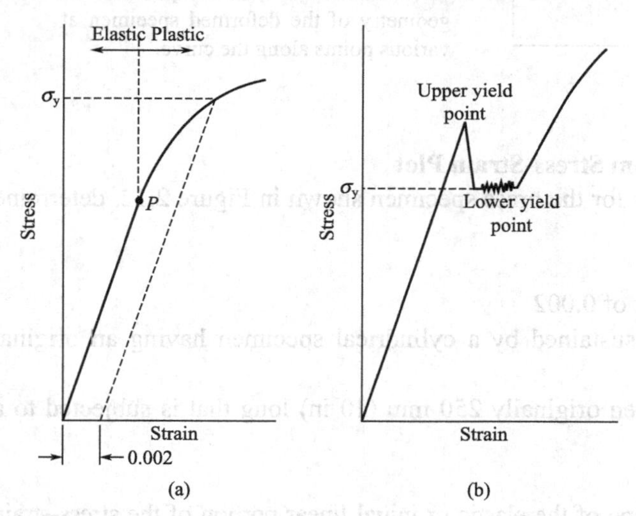

Figure 2.8 (a) Typical stress-strain behavior for a metal showing elastic and plastic deformations, the proportional limit P, and the yield strength σ_y as determined using the 0.002 strain offset method.
(b) Representative stress-strain behavior found for some steels demonstrating the yield point phenomenon.

The magnitude of the yield strength for a metal is a measure of its resistance to plastic deformation. Yield strengths may range from 35MPa (5000 psi) for a low-strength aluminum to over 1400MPa (200000 psi) for high-strength steels.

2.6.2 Tensile Strength

After yielding, the stress necessary to continue plastic deformation in metals increases to a maximum, point B in Figure 2.9, and then decreases to the eventual fracture, point k. The tensile strength σ_b (MPa or psi) is the stress at the maximum on the engineering stress-strain curve (Figure 2.9). This corresponds to the maximum stress that can be sustained by a structure in tension; if this stress is applied and maintained, fracture will result. All deformation up to this point is uniform throughout the narrow region of the tensile specimen. However, at this maximum stress, a small constriction or neck begins to form at some point, and all subsequent deformation is confined at this neck, as indicated by the schematic specimen insets in Figure 2.9. This phenomenon is termed "necking," and fracture ultimately occurs at the neck. The fracture strength corresponds to the stress at fracture.

Tensile strengths may vary anywhere from 50MPa (7000 psi) for an aluminum to as high as 3000MPa (450000 psi) for the high-strength steels. Ordinarily, when the strength of a metal is cited for design purposes, the yield strength is used. This is because by the time a stress corresponding to the tensile strength has been applied, often a structure has experienced so much plastic deformation that it is useless. Furthermore, fracture strengths are not normally specified for engineering design purposes.

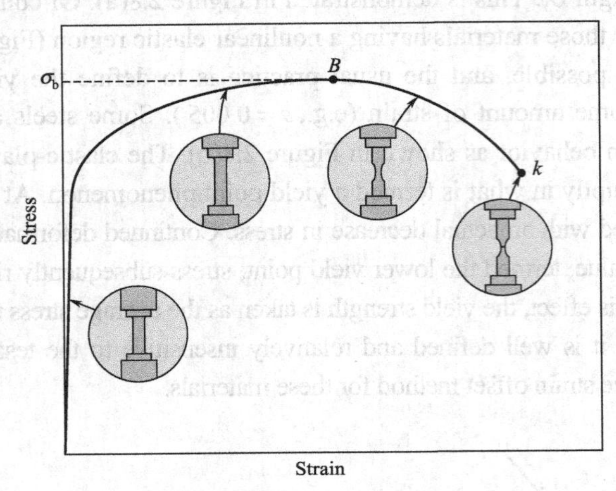

Figure 2.9 Typical engineering stress-strain behavior to fracture, point k. The tensile strength σ_b is indicated at point B. The circular insets represent the geometry of the deformed specimen at various points along the curve.

Example Problem 2.3

Mechanical Property Determinations from Stress-Strain Plot

From the tensile stress-strain behavior for the brass specimen shown in Figure 2.12, determine the following:

(1) The modulus of elasticity

(2) The yield strength at a strain offset of 0.002

(3) The maximum load that can be sustained by a cylindrical specimen having an original diameter of 12.8 mm (0.505 in)

(4) The change in length of a specimen originally 250 mm (10 in) long that is subjected to a tensile stress of 345MPa (50000 psi)

Solution

(1) The modulus of elasticity is the slope of the elastic or initial linear portion of the stress-strain curve. The strain axis has been expanded in the inset, Figure 2.12, to facilitate this computation. The slope of this linear region is the rise over the run, or the change in stress divided by the corresponding change in strain; in mathematical terms

$$E = slope = \frac{\Delta\sigma}{\Delta\varepsilon} = \frac{\sigma_2 - \sigma_1}{\varepsilon_2 - \varepsilon_1} \tag{2.10}$$

In as much as the line segment passes through the origin, it is convenient to take both σ_1 and ε_1 as zero. If σ_2 is arbitrarily taken as 150MPa, then ε_2 will have a value of 0.0016. Therefore

$$E = \frac{(150-0)\text{MPa}}{0.0016-0} = 93.8\text{GPa}(13.6\times10^6\ \text{psi})$$

which is very close to the value of 97GPa (14psi) given for brass in Table 2.1.

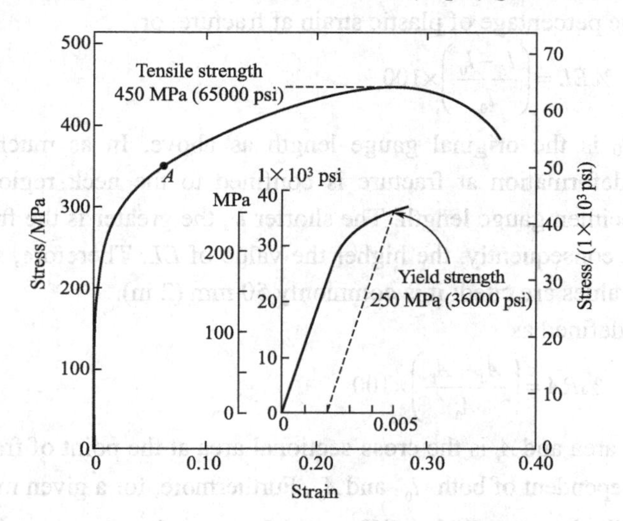

Figure 2.10 The stress-strain behavior for the brass specimen discussed in Example Problem 2.3.

(2) The 0.002 strain offset line is constructed as shown in the inset; its intersection with the stress-strain curve is at approximately 250MPa (36000psi), which is the yield strength of the brass.

(3) The maximum load that can be sustained by the specimen is calculated by using Equation 2.1, in which σ is taken to be the tensile strength, from Figure 2.10, 450MPa (65000psi). Solving for F, the maximum load, yields

$$F=\sigma A_0=\sigma\left(\frac{d_0}{2}\right)^2\pi=(450\times10^6\mathrm{N/m^2})\times\left(\frac{12.8\times10^{-3}\mathrm{m}}{2}\right)^2\pi=57900\mathrm{N}(13000\mathrm{lb_f})$$

(4) To compute the change in length, Δl, in Equation 2.2, it is first necessary to determine the strain that is produced by a stress of 345MPa. This is accomplished by locating the stress point on the stress-strain curve, point *A*, and reading the corresponding strain from the strain axis, which is approximately 0.06. Inasmuch as $l_0=250$ mm, we have

$$\Delta l=\varepsilon l_0=0.06\times250\mathrm{mm}=15\mathrm{mm}(0.6\mathrm{in})$$

2.6.3 Ductility

Ductility is another important mechanical property. It is a measure of the degree of plastic deformation that has been sustained at fracture. A material that experiences very little or no plastic deformation upon fracture is termed brittle. The tensile stress-strain behaviors for both ductile and brittle materials are schematically illustrated in Figure 2.11.

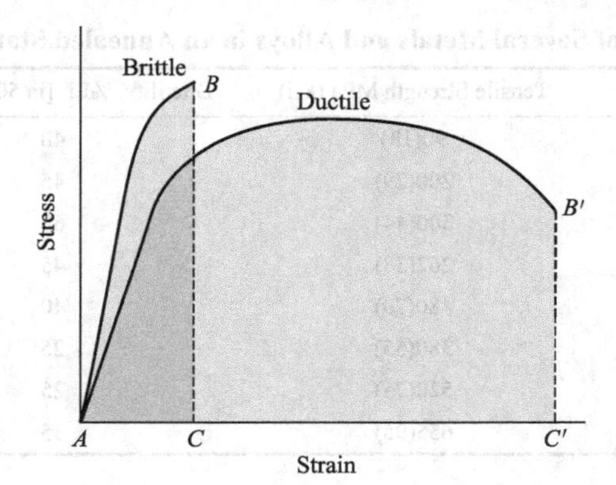

Figure 2.11 Schematic representations of tensile stress-strain behavior for brittle and ductile materials loaded to fracture.

Ductility may be expressed quantitatively as either percent elongation or percent reduction in area. The percent elongation %EL is the percentage of plastic strain at fracture, or

$$\%EL = \left(\frac{l_f - l_0}{l_0}\right) \times 100 \tag{2.11}$$

where l_f is the fracture length and l_0 is the original gauge length as above. In as much as a significant proportion of the plastic deformation at fracture is confined to the neck region, the magnitude of *%EL* will depend on specimen gauge length. The shorter l_0, the greater is the fraction of total elongation from the neck and, consequently, the higher the value of *EL*. Therefore, should be specified when percent elongation values are cited; it is commonly 50 mm (2 in).

Percent reduction in area *%RA* is defined as

$$\%RA = \left(\frac{A_0 - A_f}{A_0}\right) \times 100 \tag{2.12}$$

where A_0 is the original cross-sectional area and A_f is the cross-sectional area at the point of fracture. Percent reduction in area values are independent of both l_0 and A_0. Furthermore, for a given material the magnitudes of *%EL* and *%RA* will, in general, be different. Most metals possess at least a moderate degree of ductility at room temperature; however, some become brittle as the temperature is lowered.

Knowledge of the ductility of materials is important for at least two reasons. First, it indicates to a designer the degree to which a structure will deform plastically before fracture. Second, it specifies the degree of allowable deformation during fabrication operations. We sometimes refer to relatively ductile materials as being "forgiving," in the sense that they may experience local deformation without fracture should there be an error in the magnitude of the design stress calculation. Brittle materials are approximately considered to be those having a fracture strain of less than about 5%. Thus, several important mechanical properties of metals may be determined from tensile stress-strain tests. Table 2.2 presents some typical room-temperature values of yield strength, tensile strength, and ductility for several of the common metals. These properties are sensitive to any prior deformation, the presence of impurities, and/or any heat treatment to which the metal has been subjected. The modulus of elasticity is one mechanical parameter that is insensitive to these treatments. As with modulus of elasticity, the magnitudes of both yield and tensile strengths decline with increasing temperature; just the reverse holds for ductility—it usually increases with temperature. Figure 2.12 shows how the stress-strain behavior of iron varies with temperature.

Table 2.2 Typical Mechanical Properties of Several Metals and Alloys in an Annealed State

Metal Alloy	Yield Strength MPa (ksi)	Tensile Strength MPa (ksi)	Ductility, %EL [in 50mm (2in)]
Aluminum	35(5)	90(13)	40
Copper	69(10)	200(29)	45
Brass(70%Cu, 30%Zn)	75(11)	300(44)	68
Iron	130(19)	262(38)	45
Nickel	138(20)	480(70)	40
Steel(1020)	180(26)	380(55)	25
Titanium	450(65)	520(75)	25
Molybdenum	565(82)	655(95)	35

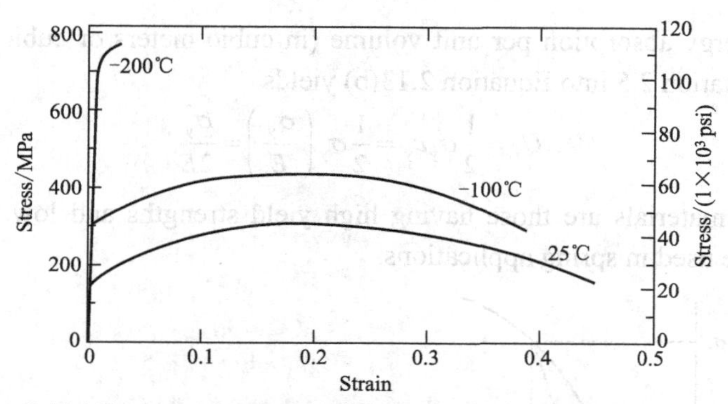

Figure 2.12 Engineering stress-strain behavior for iron at three temperatures.

2.6.4 Toughness

Toughness is a mechanical term that is used in several contexts; loosely speaking, it is a measure of the ability of a material to absorb energy up to fracture. Specimen geometry as well as the manner of load application is important in toughness determinations. For dynamic (high strain rate) loading conditions and when a notch (or point of stress concentration) is present, notch toughness is assessed by using an impact test. Furthermore, fracture toughness is a property indicative of a material's resistance to fracture when a crack is present.

For the static (low strain rate) situation, toughness may be ascertained from the results of a tensile stress-strain test. It is the area under the $\sigma-\varepsilon$ curve up to the point of fracture. The units for toughness are the same as for resilience (i.e., energy per unit volume of material). For a material to be tough it must display both strength and ductility; often, ductile materials are tougher than brittle ones.

This is demonstrated in Figure 2.13, in which the stress-strain curves are plotted for both material types. Hence, even though the brittle material has higher yield and tensile strengths, it has a lower toughness than the ductile one, by virtue of lack of ductility; this is deduced by comparing the areas *ABC* and *A'B'C'* in Figure 2.13.

2.6.5 Resilience

Resilience is the capacity of a material to absorb energy when it is deformed elastically and then, upon unloading, to have this energy recovered. The associated property is the modulus of resilience, U_r, which is the strain energy per unit volume required to stress a material from an unloaded state up to the point of yielding.Computationally, the modulus of resilience for a specimen subjected to a uniaxial tension test is just the area under the engineering stress-strain curve taken to yielding (Figure 2.13), or

$$U_r = \int_0^{\varepsilon_y} \sigma \mathrm{d}\varepsilon \tag{2.13a}$$

Assuming a linear elastic region,

$$U_r = \frac{1}{2}\sigma_y \varepsilon_y \tag{2.13b}$$

in which is the strain at yielding. The units of resilience are the product of the units from each of the two axes of the stress-strain plot. For SI units, this is joules per cubic meter (J/m³, equivalent to Pa), whereas with Customary U.S. units it is inch-pounds force per cubic inch (in.-lb$_f$/in.³, equivalent to psi). Both joules and inch-pounds force are units of energy, and thus this area under the stress–strain curve represents energy absorption per unit volume (in cubic meters or cubic inches) of material. Incorporation of Equation 2.5 into Equation 2.13(b) yields

$$U_r = \frac{1}{2}\sigma_y \varepsilon_y = \frac{1}{2}\sigma_y\left(\frac{\sigma_y}{E}\right) = \frac{\sigma_y^2}{2E} \tag{2.14}$$

Thus, resilient materials are those having high yield strengths and low moduli of elasticity; such alloys would be used in spring applications.

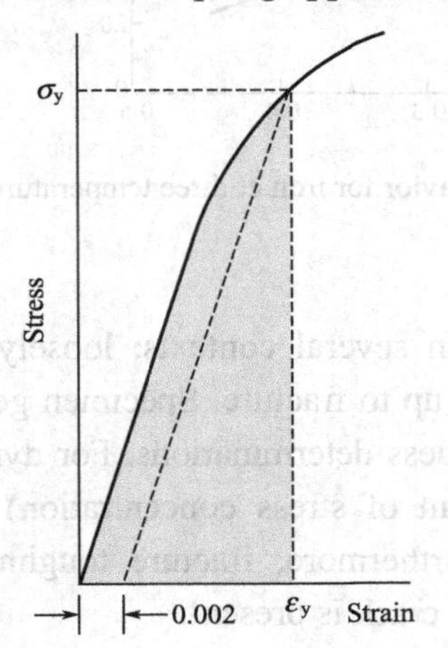

Figure 2.13 Schematic representation showing how modulus of resilience (corresponding to the shaded area) is determined from the tensile stress-strain behavior of a material.

2.7 True Stress And Strain

From Figure 2.9, the decline in the stress necessary to continue deformation past the maximum, point *B*, seems to indicate that the metal is becoming weaker. This is not at all the case; as a matter of fact, it is increasing in strength. However, the cross-sectional area is decreasing rapidly within the neck region, where deformation is occurring. This results in a reduction in the load-bearing capacity of the specimen. The stress, as computed from Equation 2.1, is on the basis of the original cross-sectional area before any deformation, and does not take into account this reduction in area at the neck.

Sometimes it is more meaningful to use a true stress-true strain scheme. True stress σ_T is defined as the load *F* divided by the instantaneous cross-sectional area over which deformation is occurring (i.e., the neck, past the tensile point), or

$$\sigma_T = \frac{F}{A_i} \tag{2.15}$$

As the applied force *F* increases, so does the length of the specimen. For an increase d*F*, the length *l* increases by d*l*. The normalized (per unit length) increase in length is equal to

$$d\varepsilon = \frac{dl}{l}$$

or, upon integration,

$$\varepsilon = \int_{l_0}^{l_1} \frac{dl}{l} = \ln\frac{l_1}{l_0} \tag{2.16}$$

where l_0 is the original length. Furthermore, it is occasionally more convenient to represent strain as true strain ε_T defined by

$$\varepsilon_T = \ln\frac{l_i}{l_0} \tag{2.17}$$

If no volume change occurs during deformation—that is, if

$$A_i l_i = A_0 l_0$$

true and engineering stress and strain are related according to

$$\sigma_T = \sigma(1+\varepsilon) \tag{2.18a}$$

$$\varepsilon_T = \ln(1+\varepsilon) \tag{2.18b}$$

Equation 2.18a and Equation 2.18b are valid only to the onset of necking; beyond this point true stress and strain should be computed from actual load, cross-sectional area, and gauge length measurements.

A schematic comparison of engineering and true stress-strain behaviors is made in Figure 2.14. It is worth noting that the true stress necessary to sustain increasing strain continues to rise past the tensile point B'.

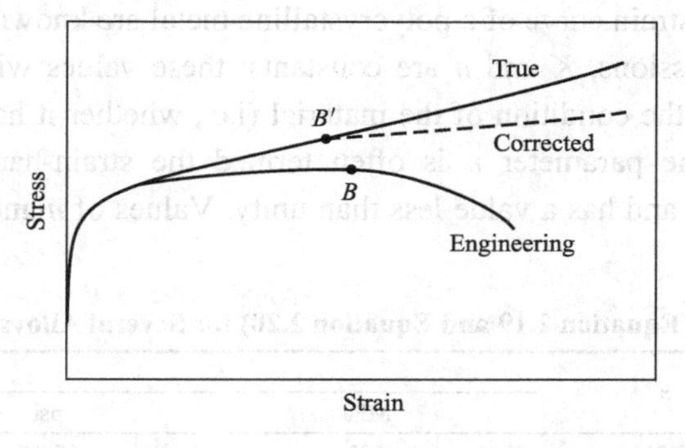

Figure 2.14 A comparison of typical tensile engineering stress-strain and true stress-strain behaviors. Necking begins at point B on the engineering curve, which corresponds to B' on the true curve. The "corrected" true stress-strain curve takes into account the complex stress state within the neck region.

Coincident with the formation of a neck is the introduction of a complex stress state within the neck region (i.e., the existence of other stress components in addition to the axial stress). As a consequence, the correct stress (axial) within the neck is slightly lower than the stress computed from the applied load and neck crosssectional area. This leads to the "corrected" curve in Figure 2.15.

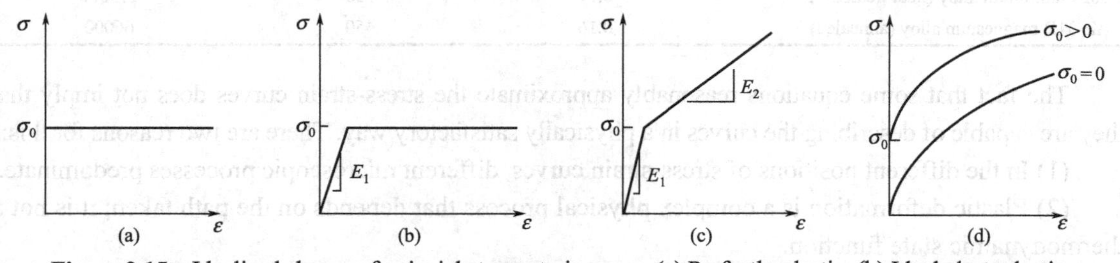

Figure 2.15 Idealized shapes of uniaxial stress-strain curve. (a) Perfectly plastic. (b) Ideal elastoplastic. (c) Ideal elastoplastic with linear work-hardening. (d) Parabolic work-hardening ($\sigma_T = \sigma_0 + K\varepsilon_T^n$).

All of the preceding curves, as well as other ones, are represented schematically by simple equations in various ways. Figure 2.15 shows four different idealized shapes for stress-strain curves. Note that these are true-stress-true-strain curves. When we have a large amount of plastic deformation, the plastic strain is large with respect to the elastic strain, and the latter can be neglected. If the material does not work-harden, the plastic curve is horizontal, and the idealized behavior is called perfectly plastic. This is shown in Figure 2.15(a). If the plastic deformation is not so large, the elastic portion of the curve cannot be neglected, and one has an ideal elastoplastic material [Figure 2.15(b)]. A further approximation to the behavior of real materials is the ideal elastoplastic behavior depicted in Figure 2.15(c); this is a linear curve with two slopes E_1 and E_2 that represent the material's elastic and plastic behavior, respectively. One could represent the behavior of the steels in Figure 2.3 fairly well by this elastoplastic, linear work-hardening behavior. It can be seen that $E_2 << E_1$. For example, for annealed steel, $E_2 \approx 70$MPa, while $E_1 = 210$GPa. However, a better representation of the work-hardening behavior is obtained by assuming a gradual decrease in the slope of the curve as plastic deformation proceeds [shown in Figure 2.15(d)].

For some metals and alloys the region of the true stress-strain curve from the onset of plastic deformation to the point at which necking begins may be approximated by

$$\sigma_T = K\varepsilon_T^n \tag{2.19}$$

Where $n<1$. This response is usually called ''parabolic'' hardening, and one can translate it upward by assuming a yield stress σ_0, so that Equation 2.19 becomes

$$\sigma_T = \sigma_0 + K\varepsilon_T^n \tag{2.20}$$

These equations that describe the stress-strain curve of a polycrystalline metal are known as the Ludwik-Hollomon equations. In these expressions, K and n are constants; these values will vary from alloy to alloy, and will also depend on the condition of the material (i.e., whether it has been plastically deformed, heat treated, etc.). The parameter n is often termed the strain-hardening exponent (or the work-hardening coefficient) and has a value less than unity. Values of n and K for several alloys are contained in Table 2.3.

Table 2.3 Tabulation of n and K Values (Equation 2.19 and Equation 2.20) for Several Alloys

Material	n	K	
		MPa	psi
Low-carbon steel (annealed)	0.21	600	87000
4340 steel alloy (tempered at 315℃)	0.12	2650	385000
304 stainless steel (annealed)	0.44	1400	205000
Copper (annealed)	0.44	530	76500
Naval brass (annealed)	0.21	585	85000
2024 aluminum alloy (heat treated-T3)	0.17	780	113000
AZ-31B magnesium alloy (annealed)	0.16	450	66000

The fact that some equations reasonably approximate the stress-strain curves does not imply that they are capable of describing the curves in a physically satisfactory way. There are two reasons for this:

(1) In the different positions of stress-strain curves, different microscopic processes predominate.

(2) Plastic deformation is a complex physical process that depends on the path taken; it is not a thermodynamic state function.

That is to say, the accumulated plastic deformation is not uniquely related to the dislocation structure of the material. This being so, it is not very likely that simple expressions could be derived for the stress-strain curves in which the parameters would have definite physical significance.

Some alloys, such as stainless steels, undergo martensitic phase transformations induced by plastic strain. This type of transformation alters the stress-strain curve. Other alloys undergo mechanical twinning beyond a threshold stress (or strain), which affects the shape of the curve. In these cases, it is necessary to divide the plastic regime into stages. It is often useful to plot the slope of the stress-strain curve vs. stress (or strain) to reveal changes in mechanism more clearly.

In spite of its limitations, the Ludwik-Hollomon Equation 2.21 is the most common representation of plastic response. When n=0, it represents ideal plastic behavior (no work-hardening). More general forms of this equation, incorporating both strain rate and thermal effects, are often used to represent the response of metals; in that case they are called constitutive equations. The flow stress of metals increases with increasing strain rate and decreasing temperature, because thermally activated dislocation motion is inhibited.

The Johnson-Cook equation

$$\sigma = (\sigma_0 + K\varepsilon^n)\left(1 + C\ln\frac{\dot{\varepsilon}}{\dot{\varepsilon}_0}\right)\left[1 - \left(\frac{T - T_r}{T_m - T_r}\right)^m\right] \tag{2.21}$$

is widely used in large-scale deformation codes. The three groups of terms in parentheses represent work-hardening, strain rate, and thermal effects, respectively. The constants K, n, C, and m are material parameters, and T_r is the reference temperature, T_m the melting point, and $\dot{\varepsilon}_0$ the reference strain rate. There are additional equations that incorporate the microstructural elements such as grain size and dislocation interactions and dynamics: they are therefore called "physically based." The effects of these factors are very complex and cannot be simply "plugged into" the equations.

Example Problem 2.4

Ductility and True-Stress At Fracture Computations

A cylindrical specimen of steel having an original diameter of 12.8 mm (0.505 in) is tensile tested to fracture and found to have an engineering fracture strength σ_f of 460MPa (67000 psi). If its cross-sectional diameter at fracture is 10.7 mm (0.422 in), determine:

(1) The ductility in terms of percent reduction in area

(2) The true stress at fracture

Solution

(1) Ductility is computed using Equation 2.12, as

$$\%RA = \frac{\left(\frac{12.8\text{mm}}{2}\right)^2\pi - \left(\frac{10.7\text{mm}}{2}\right)^2\pi}{\left(\frac{12.8\text{mm}}{2}\right)^2\pi} \times 100 = \frac{128.7 - 89.9}{128.7} \times 100 = 30\%$$

(2) True stress is defined by Equation 2.15, where in this case the area is taken as the fracture area A_f. However, the load at fracture must first be computed from the fracture strength as

$$F = \sigma_f A_0 = (460 \times 10^6\text{N/m}^2) \times (128.7\text{mm}^2) \times \left(\frac{1\text{m}^2}{10^6\text{mm}^2}\right) = 59200\text{N}$$

Thus, the true stress is calculated as

$$\sigma_T = \frac{F}{A_f} = \frac{59200\text{N}}{89.9\text{mm}^2 \times \left(\frac{1\text{m}^2}{10^6\text{mm}^2}\right)} = 6.6 \times 10^8\text{N/mm}^2 = 660\text{MPa}(95700\text{psi})$$

Example Problem 2.5

Calculation of Strain-Hardening Exponent

Compute the strain-hardening exponent n in Equation 2.19 for an alloy in which a true stress of 415 MPa (60000psi) produces a true strain of 0.10; assume a value of 1035 MPa (150000psi) for K.

Solution

This requires some algebraic manipulation of Equation 2.19 so that n becomes the dependent parameter. This is accomplished by taking logarithms and rearranging. Solving for n yields

$$n = \frac{\lg\sigma_T - \lg K}{\lg\varepsilon_T} = \frac{\lg(415\text{MPa}) - \lg(1035\text{MPa})}{\lg(0.1)} = 0.4$$

2.8 Elastic Recovery After Plastic Deformation

Upon release of the load during the course of a stress-strain test, some fraction of the total deformation is recovered as elastic strain. This behavior is demonstrated in Figure 2.18, a schematic engineering stress-strain plot. During the unloading cycle, the curve traces a near straight-line path from the point of unloading (point *D*), and its slope is virtually identical to the modulus of elasticity, or parallel to the initial elastic portion of the curve. The magnitude of this elastic strain, which is regained during unloading, corresponds to the strain recovery, as shown in Figure 2.16. If the load is reapplied, the curve will traverse essentially the same linear portion in the direction opposite to unloading; yielding will again occur at the unloading stress level where the unloading began. There will also be an elastic strain recovery associated with fracture.

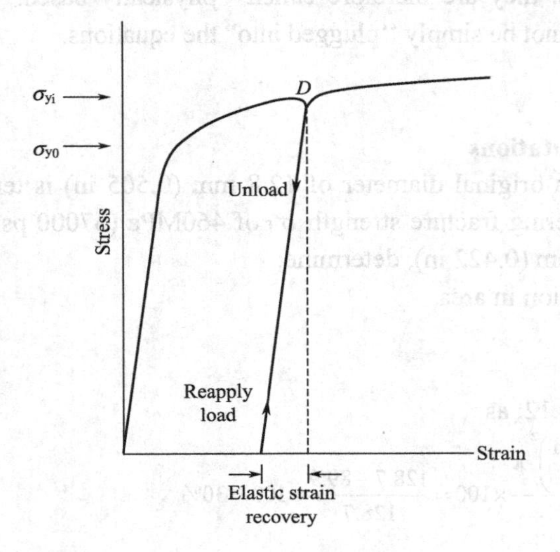

Figure 2.16 Schematic tensile stress-strain diagram showing the phenomena of elastic strain recovery and strain hardening. The initial yield strength is designated as is σ_{y0} . σ_{yi} the yield strength after releasing the load at point *D*, and then upon reloading.

2.9 Compressive, Shear, And Torsion Deformation

Of course, metals may experience plastic deformation under the influence of applied compressive, shear, and torsional loads. The resulting stress-strain behavior into the plastic region will be similar to the tensile counterpart (Figure 2.10: yielding and the associated curvature). However, for compression, there will be no maximum, since necking does not occur; furthermore, the mode of fracture will be different from that for tension.

2.10 Hardness

Another mechanical property that may be important to consider is hardness, which is a measure of a material's resistance to localized plastic deformation (e.g., a small dent or a scratch). Early hardness tests were based on natural minerals with a scale constructed solely on the ability of one material to scratch another that was softer. A qualitative and somewhat arbitrary hardness indexing scheme was devised, termed the Mohs scale, which ranged from 1 on the soft end for talc to 10 for diamond. Quantitative hardness techniques have been developed over the years in which a small indenter is forced into the surface of a material to be tested, under controlled conditions of load and rate of application. The depth or size of the resulting indentation is measured, which in turn is related to a hardness number; the softer the material, the larger and deeper is the indentation, and the lower the hardness index number. Measured hardnesses are only relative (rather than absolute), and care should be exercised when comparing values determined by different techniques.

Hardness tests are performed more frequently than any other mechanical test for several reasons:

(1) They are simple and inexpensive—ordinarily no special specimen need be prepared, and the testing apparatus is relatively inexpensive.

(2) The test is nondestructive—the specimen is neither fractured nor excessively deformed; a small indentation is the only deformation.

(3) Other mechanical properties often may be estimated from hardness data, such as tensile strength.

2.10.1 Brinell Hardness Tests

In Brinell tests, as in Rockwell measurements, a hard, spherical indenter is forced into the surface of the metal to be tested. The diameter of the hardened steel (or tungsten carbide) indenter is 10.00mm (0.394 in). Standard loads range between 500kg and 3000kg in 500kg increments; during a test, the load is maintained constant for a specified time (between 10s and 30s). Harder materials require greater applied loads. The Brinell hardness number, HB, is a function of both the magnitude of the load and the diameter of the resulting indentation (see Table 2.4). This diameter is measured with a special low-power microscope, utilizing a scale that is etched on the eyepiece. The measured diameter is then converted to the appropriate HB number using a chart; only one scale is employed with this technique.

Semiautomatic techniques for measuring Brinell hardness are available. These employ optical scanning systems consisting of a digital camera mounted on a flexible probe, which allows positioning of the camera over the indentation. Data from the camera are transferred to a computer that analyzes the indentation, determines its size, and then calculates the Brinell hardness number. For this technique, surface finish requirements are normally more stringent that for manual measurements.

2.10.2 Rockwell Hardness Tests

The Rockwell tests constitute the most common method used to measure hardness because they are so simple to perform and require no special skills. Several different scales may be utilized from possible combinations of various indenters and different loads, which permit the testing of virtually all metal alloys (as well as some polymers). Indenters include spherical and hardened steel balls having diameters of 1/16in, 1/8in, 1/4in and 1/2in (1.588mm, 3.175mm, 6.350mm, and 12.70mm), and a conical diamond (Brale) indenter, which is used for the hardest materials.

With this system, a hardness number is determined by the difference in depth of penetration resulting from the application of an initial minor load followed by a larger major load; utilization of a minor load enhances test accuracy. On the basis of the magnitude of both major and minor loads, there are two types of tests: Rockwell and superficial Rockwell. For Rockwell, the minor load is 10 kg, whereas major loads are 60kg, 100kg, and 150kg. Each scale is represented by a letter of the alphabet; several are listed with the corresponding indenter and load in Table 2.4 and Table 2.5a.

For superficial tests, 3kg is the minor load; 15kg, 30kg, and 45kg are the possible major load values. These scales are identified by a 15, 30, or 45 (according to load), followed by N, T, W, X, or Y, depending on indenter. Superficial tests are frequently performed on thin specimens. Table 2.5b presents several superficial scales.

When specifying Rockwell and superficial hardnesses, both hardness number and scale symbol must be indicated. The scale is designated by the symbol HR followed by the appropriate scale identification. For example, 80HRB represents a Rockwell hardness of 80 on the B scale, and 60HR30W indicates a superficial hardness of 60 on the 30W scale.

For each scale, hardnesses may range up to 130; however, as hardness values rise above 100 or drop below 20 on any scale, they become inaccurate; and because the scales have some overlap, in such a situation it is best to utilize the next harder or softer scale.

Inaccuracies also result if the test specimen is too thin, if an indentation is made too near a specimen edge, or if two indentations are made too close to one another. Specimen thickness should be at least ten times the indentation depth, whereas allowance should be made for at least three indentation diameters between the center of one indentation and the specimen edge, or to the center of a second indentation. Furthermore, testing of specimens stacked one on top of another is not recommended. Also, accuracy is dependent on the indentation being made into a smooth flat surface. The modern apparatus for making Rockwell hardness measurements (see the chapter-opening photograph for this chapter) is automated and very simple to use; hardness is read directly, and each measurement requires only a few seconds.

The modern testing apparatus also permits a variation in the time of load application. This variable must also be considered in interpreting hardness data. Maximum specimen thickness as well as indentation position (relative to specimen edges) and minimum indentation spacing requirements are the same as for Brinell tests. In addition, a well- defined indentation is required; this necessitates a smooth flat surface in which the indentation is made.

Table 2.4 Hardness-Testing Techniques

Test	Indenter	Shape of Indentation: Side View	Shape of Indentation: Top View	Load	Formula for Hardness Number①
Brinell	10-mm sphere of steel or tungsten carbide	D; d	d	P	$HB=\frac{2P}{\pi D[D-\sqrt{D^2-d^2}]}$
Vickers microhardness	Diamond pyramid	136°	d_1, d_1	P	$HV=1.854P/d_1^2$
Knoop microhardness	Diamond pyramid	$l/b=7.11$; $b/t=4.00$; t	b; l	P	$HK=14.2P/l^2$
Rockwell and Superficial Rockwell	Diamond cone; $\frac{1}{16}$in, $\frac{1}{8}$in, $\frac{1}{4}$in, $\frac{1}{2}$in diameter steel spheres	120°		60kg, 100kg, 150kg: Rockwell; 15kg, 30kg, 45kg: Superficial Rockwell	

① For the hardness formulas given, P(the applied load) is in kg, while D, d, d_1, and l are all in mm.

Source: Adapted from H.W. Hayden, W.G. Moffatt, and J. Wulff, *The Structure and Properties* of *Materials*, Vol.III, *Mechanical Behavior*. Copyright © 1965 by John Wiley & Sons, New York. Reprinted by permission of John Wiley & Sons, Inc.

Table 2.5a Rockwell Hardness Scales

Scale Symbol	Indenter	Major Load/kg
A	Diamond	60
B	$\frac{1}{16}$-in.ball	100
C	Diamond	150
D	Diamond	100
E	$\frac{1}{8}$-in.ball	100
F	$\frac{1}{16}$-in.ball	60
G	$\frac{1}{16}$-in.ball	150
H	$\frac{1}{8}$-in.ball	60
K	$\frac{1}{8}$-in.ball	150

Table 2.5b Superficial Rockwell Hardness Scales

Scale Symbol	Indenter	Major Load/kg
15N	Diamond	15
30N	Diamond	30
45N	Diamond	45
15T	$\frac{1}{16}$-in.ball	15
30T	$\frac{1}{16}$-in.ball	30
45T	$\frac{1}{16}$-in.ball	45
15W	$\frac{1}{8}$-in.ball	15
30W	$\frac{1}{8}$-in.ball	30
45W	$\frac{1}{8}$-in.ball	45

2.10.3 Knoop and Vickers Microindentation Hardness Tests

Two other hardness-testing techniques are Knoop and Vickers (sometimes also called diamond pyramid). For each test a very small diamond indenter having pyramidal geometry is forced into the surface of the specimen. Applied loads are much smaller than for Rockwell and Brinell, ranging between 1g and 1000g. The resulting impression is observed under a microscope and measured; this measurement is then converted into a hardness number (Table 2.4). Careful specimen surface preparation (grinding and polishing) may be necessary to ensure a well-defined indentation that may be accurately measured. The Knoop and Vickers hardness numbers are designated by *HK* and *HV*, respectively, and hardness scales for both techniques are approximately equivalent. Knoop and Vickers are referred to as microindentation-testing methods on the basis of indenter size. Both are well suited for measuring the hardness of small, selected specimen regions; furthermore, Knoop is used for testing brittle materials such as ceramics.

The modern microindentation hardness-testing equipment has been automated by coupling the indenter apparatus to an image analyzer that incorporates a computer and software package. The software controls important system functions to include indent location, indent spacing, computation of hardness values, and plotting of data. Other hardness-testing techniques are frequently employed but will not be discussed here; these include ultrasonic microhardness, dynamic (Scleroscope), durometer (for plastic and elastomeric materials), and scratch hardness tests. These are described in references provided at the end of the chapter.

2.10.4 Correlation Between Hardness and Tensile Strength

Both tensile strength and hardness are indicators of a metal's resistance to plastic deformation. Consequently, they are roughly proportional, as shown in Figure 2.17, for tensile strength as a function of the *HB* for cast iron, steel, and brass. The same proportionality relationship does not hold for all metals, as Figure 2.19 indicates. As a rule of thumb for most steels, the *HB* and the tensile strength are related according to

$$TS(\text{MPa}) = 3.45 \times HB \tag{2.22a}$$

$$TS(\text{psi}) = 500 \times HB \tag{2.22b}$$

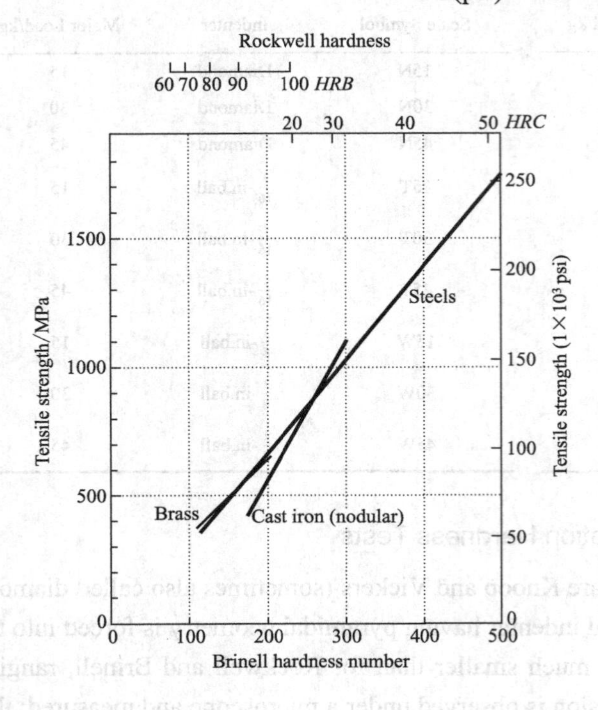

Figure 2.17 Relationships between hardness and tensile strength for steel, brass, and cast iron. [Data taken from Metals Handbook: Properties and Selection: Irons and Steels, Vol. 1, 9th edition, B. Bardes (Editor), American Society for Metals, 1978, 36 and 461; and Metals Handbook: Properties and Selection: Nonferrous Alloys and Pure Metals, Vol. 2, 9th edition, H. Baker (Managing Editor), American Society for Metals, 1979, 327.]

SUMMARY

Concepts of Stress and Strain

Stress-Strain Behavior

Elastic Properties of Materials

True Stress and Strain

A number of the important mechanical properties of materials, predominantly metals, have been discussed in this chapter. Concepts of stress and strain were first introduced. Stress is a measure of an applied mechanical load or force, normalized to take into account cross-sectional area. Two different stress parameters were defined—engineering stress and true stress. Strain represents the amount of deformation induced by a stress; both engineering and true strains are used.

Some of the mechanical characteristics of metals can be ascertained by simple stress-strain tests. There are four test types: tension, compression, torsion, and shear. Tensile are the most common. A material that is stressed first undergoes elastic, or nonpermanent, deformation, wherein stress and strain are proportional. The constant of proportionality is the modulus of elasticity for tension and compression, and is the shear modulus when the stress is shear. Poisson's ratio represents the negative ratio of transverse and longitudinal strains.

Tensile Properties

The phenomenon of yielding occurs at the onset of plastic or permanent deformation; yield strength is determined by a strain offset method from the stress-strain behavior, which is indicative of the stress at which plastic deformation begins. Tensile strength corresponds to the maximum tensile stress that may be sustained by a specimen, whereas percents elongation and reduction in area are measures of ductility—the amount of plastic deformation that has occurred at fracture.

Resilience is the capacity of a material to absorb energy during elastic deformation; modulus of resilience is the area beneath the engineering stress-strain curve up to the yield point. Also, static toughness represents the energy absorbed during the fracture of a material, and is taken as the area under the entire engineering stress-strain curve. Ductile materials are normally tougher than brittle ones.

Hardness

Hardness is a measure of the resistance to localized plastic deformation. In several popular hardness-testing techniques (Rockwell, Brinell, Knoop, and Vickers) a small indenter is forced into the surface of the material, and an index number is determined on the basis of the size or depth of the resulting indentation. For many metals, hardness and tensile strength are approximately proportional to each other.

IMPORTANT TERMS AND CONCEPTS

Anelasticity	Design stress	Ductility
Elastic deformation	Elastic recovery	Engineering strain
Engineering stress	Hardness	Modulus of elasticity
Plastic deformation	Poisson's ratio	Proportional limit
Resilience	Safe stress	Shear
Tensile strength	Toughness	True strain
True stress	Yielding	Yield strength

REFERENCES

1. ASM Handbook, Vol. 8, Mechanical Testing and Evaluation, ASM International, Materials Park, OH, 2000
2. Boyer, H. E. (Editor), Atlas of Stress-Strain Curves, 2nd edition,ASM International, Materials Park, OH, 2002
3. Chandler, H. (Editor), Hardness Testing, 2nd edition, ASM International, Materials Park, OH, 2000
4. Courtney, T. H., Mechanical Behavior of Materials, 2nd edition, McGraw-Hill Higher Education, Burr Ridge, IL, 2000
5. Davis, J. R. (Editor), Tensile Testing, 2nd edition, ASM International, Materials Park, OH, 2004
6. Dieter, G. E., Mechanical Metallurgy, 3rd edition, McGraw-Hill Book Company, New York, 1986
7. Dowling, N. E., Mechanical Behavior of Materials, 2nd edition, Prentice Hall PTR, Paramus, NJ, 1998
8. McClintock, F. A. and A. S. Argon, Mechanical Behavior of Materials, Addison-Wesley Publishing Co., Reading, MA, 1966 Reprinted by CBLS Publishers, Marietta, OH, 1993
9. Meyers, M. A. and K. K. Chawla, Mechanical Behavior of Materials, Prentice Hall PTR, Paramus, NJ, 1999
10. P. Ludwik, Elemente der Technologischen Mechanik (Berlin: Springer, 1909), 32
11. J. H. Hollomon, Trans. AIME, 162 (1945) 268

QUESTIONS AND PROBLEM

2.1 Using mechanics of materials principles (i.e., equations of mechanical equilibrium applied to a free-body diagram), derive Equations 2.4a and 2.4b.

2.2 (a) Equations 2.4a and 2.4b are expressions for normal σ' and shear τ' stresses, respectively, as a function of the applied tensile stress σ and the inclination angle of the plane on which these stresses are taken (θ of Figure 2.4). Make a plot on which is presented the orientation parameters of these expressions (i.e., $\cos^2\theta$ and $\sin\theta \cos\theta$) versus θ.

(b) From this plot, at what angle of inclination is the normal stress a maximum?

(c) Also, at what inclination angle is the shear stress a maximum?

2.3 A specimen of copper having a rectangular cross section 15.2 mm×19.1 mm (0.60 in×0.75 in) is pulled in tension with 44500N (10000 lb_f) force, producing only elastic deformation. Calculate the resulting strain.

2.4 A cylindrical specimen of a nickel alloy having an elastic modulus of 207 GPa (30×10^6psi) and an original diameter of 10.2 mm (0.40 in) will experience only elastic deformation when a tensile load of 8900 N (2000 lb_f)

is applied. Compute the maximum length of the specimen before deformation if the maximum allowable elongation is 0.25 mm (0.010 in).

2.5 An aluminum bar 125 mm (5.0 in) long and having a square cross section 16.5 mm (0.65 in) on an edge is pulled in tension with a load of 66700N (15000 lb_f), and experiences an elongation of 0.43 mm (1.7×10^{-2} in). Assuming that the deformation is entirely elastic, calculate the modulus of elasticity of the aluminum.

2.6 Consider a cylindrical nickel wire 2.0 mm (0.08 in) in diameter and 3×10^4mm (1200 in) long. Calculate its elongation when a load of 300 N (67 lb_f) is applied. Assume that the deformation is totally elastic.

2.7 For a brass alloy, the stress at which plastic deformation begins is 345 MPa (50000 psi), and the modulus of elasticity is 103 GPa (15.0×10^6psi).

(a) What is the maximum load that may be applied to a specimen with a cross-sectional area of 130 mm^2 (0.2 in^2) without plastic deformation?

(b) If the original specimen length is 76 mm (3.0 in), what is the maximum length to which it may be stretched without causing plastic deformation?

2.8 A cylindrical rod of steel (E=207GPa, 30×10^6psi) having a yield strength of 310 MPa (45000psi) is to be subjected to a load of 11100N (2500 lb_f). If the length of the rod is 500mm (20.0 in), what must be the diameter to allow an elongation of 0.38 mm (0.015 in)?

2.9 Consider a cylindrical specimen of a steel alloy (Figure 2.18) 8.5 mm (0.33 in) in diameter and 80 mm (3.15 in) long that is pulled intension. Determine its elongation when a load of 65250 N (14500 lb_f) is applied.

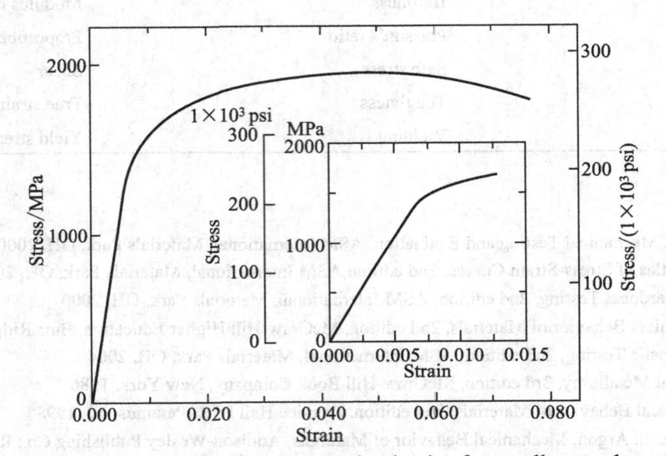

Figure 2.18 Tensile stress-strain ehavior for an alloy steel.

2.10 Figure 2.19 shows, for a gray cast iron, the tensile engineering stress-strain curve in the elastic region. Determine (a) the tangent modulus at 25 MPa (3625 psi), and (b) the secant modulus taken to 35 MPa (5000 psi).

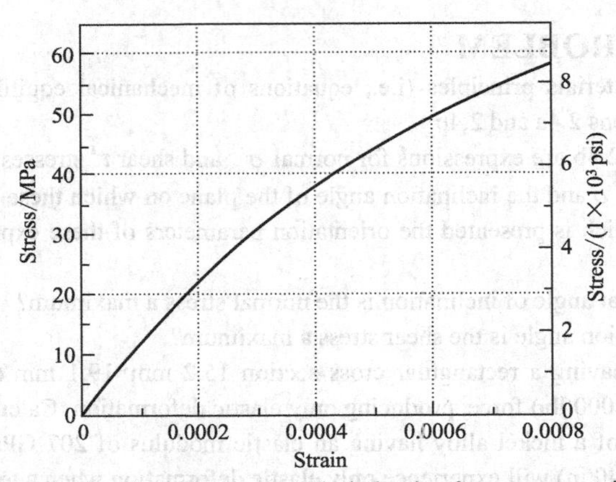

Figure 2.19 Tensile stress-strain behavior for a gray cast iron.

2.11 For single crystals of some substances, the physical properties are anisotropic; that is, they are dependent on crystallographic direction. One such property is the modulus of elasticity. For cubic single crystals, the modulus of elasticity in a general [*uvw*] direction, E_{uvw}, is described by the relationship

$$\frac{1}{E_{\langle uvw \rangle}}=\frac{1}{E_{\langle 100 \rangle}}-3\left(\frac{1}{E_{\langle 100 \rangle}}-\frac{1}{E_{\langle 111 \rangle}}\right)(\alpha^2\beta^2+\beta^2\gamma^2+\gamma^2\alpha^2)$$

where $E_{\langle 100 \rangle}$ and $E_{\langle 111 \rangle}$ are the moduli of elasticity in [100] and [111] directions, respectively; α, β and γ are the cosines of the angles between [*uvw*] and the respective [100], [010] and [001] directions. Calculate the $E_{\langle 110 \rangle}$ values for aluminum, copper, and iron.

2.12 As we know, the net bonding energy E_N between two isolated positive and negative ions is a function of interionic distance r as follows:

$$E_N=-\frac{A}{r}+\frac{B}{r^n} \tag{2.23}$$

where *A*, *B*, and *n* are constants for the particular ion pair. Equation 2.25 is also valid for the bonding energy between adjacent ions in solid materials. The modulus of elasticity *E* is proportional to the slope of the interionic force-separation curve at the equilibrium interionic separation; that is

$$E \propto \left(\frac{dF}{dr}\right)_{r_0}$$

Derive an expression for the dependence of the modulus of elasticity on these *A*, *B*, and *n* parameters (for the two-ion system) using the following procedure:

(1) Establish a relationship for the force *F* as a function of *r*, realizing that

$$F=\frac{dE_N}{dr}$$

(2) Now take the derivative dF/dr.

(3) Develop an expression for r_0, the equilibrium separation. Since r_0 corresponds to the value of *r* at the minimum of the E_N-versus-*r* curve, take the derivative dE_N/dr, set it equal to zero, and solve for *r*, which corresponds to r_0.

(4) Finally, substitute this expression for r_0 into the relationship obtained by taking dF/dr.

2.13 Using the solution to Problem 2.12, rank the magnitudes of the moduli of elasticity for the following hypothetical *X*, *Y*, and *Z* materials from the greatest to the least. The appropriate *A*, *B*, and *n* parameters (Equation 2.25) for these three materials are tabulated below; they yield E_N in units of electron volts and *r* in nanometers:

Material	*A*	*B*	*n*
X	1.5	7.0×10^{-6}	8
Y	2.0	1.0×10^{-5}	9
Z	3.5	4.0×10^{-6}	7

2.14 A cylindrical specimen of steel having a diameter of 15.2 mm (0.60 in) and length of 250 mm (10.0 in) is deformed elastically in tension with a force of 48900 N (11000lb$_f$). Using the data contained in Table 2.1, determine the following:

(1) The amount by which this specimen will elongate in the direction of the applied stress.

(2) The change in diameter of the specimen. Will the diameter increase or decrease?

2.15 A cylindrical bar of aluminum 19 mm (0.75 in) in diameter is to be deformed elastically by application of a force along the bar axis. Using the data in Table 6.1, determine the force that will produce an elastic reduction of 2.5×10^{-3} mm (1.0×10^{-4} in) in the diameter.

2.16 A cylindrical specimen of some metal alloy 10 mm (0.4 in) in diameter is stressed elastically in tension. A force of 15000 N (3370 lb$_f$) produces a reduction in specimen diameter 7×10^{-3} of mm (2.8×10^{-4} in). Compute Poisson's ratio for this material if its elastic modulus is 100 GPa (14.5×10^6 psi).

2.17 A cylindrical specimen of a hypothetical metal alloy is stressed in compression. If its original and final diameters are 30.00mm and 30.04 mm, respectively, and its final length is 105.20 mm, compute its original length if the deformation is totally elastic. The elastic and shear moduli for this alloy are 65.5 GPa and 25.4 GPa, respectively.

2.18 Consider a cylindrical specimen of some hypothetical metal alloy that has a diameter of 10.0 mm (0.39 in). A tensile force of 1500 N (340 lbf) produces an elastic reduction in diameter of 6.7×10^{-4} mm (2.64×10^{-5}in). Compute the elastic modulus of this alloy, given that Poisson's ratio is 0.35.

2.19 A brass alloy is known to have a yield strength of 240 MPa (35000 psi), a tensile strength of 310 MPa (45000 psi), and an elastic modulus of 110 GPa (16.0×10^{6} psi). A cylindrical specimen of this alloy 15.2 mm (0.60 in) in diameter and 380 mm (15.0 in) long is stressed in tension and found to elongate 1.9 mm (0.075 in). On the basis of the information given, is it possible to compute the magnitude of the load that is necessary to produce this change in length? If so, calculate the load. If not, explain why.

2.20 A cylindrical metal specimen 15.0 mm (0.59 in) in diameter and 150 mm (5.9 in) long is to be subjected to a tensile stress of 50 MPa (7250 psi); at this stress level the resulting deformation will be totally elastic.

(1) If the elongation must be less than 0.072 mm (2.83×10^{-3} in), which of the metals in Table 2.1 are suitable candidates? Why?

(2) If, in addition, the maximum permissible diameter decrease is 2.3×10^{-3}mm (9.1×10^{-5} in) when the tensile stress of 50 MPa is applied, which of the metals that satisfy the criterion in part (a) are suitable candidates? Why?

2.21 Consider the brass alloy for which the stress–strain behavior is shown in Figure 2.12. A cylindrical specimen of this material 10.0 mm (0.39 in) in diameter and 101.6 mm (4.0 in) long is pulled in tension with a force of 10,000 N (2250 lbf). If it is known that this alloy has a value for Poisson's ratio of 0.35, compute:

(1) the specimen elongation.

(2) the reduction in specimen diameter.

2.22 A cylindrical rod 120 mm long and having a diameter of 15.0 mm is to be deformed using a tensile load of 35000 N. It must not experience either plastic deformation or a diameter reduction of more than 1.2×10^{-2} mm. Of the materials listed below, which are possible candidates? Justify your choice(s).

Material	Modulus of Elasticity/GPa	Yield Strength/MPa	Poisson's Ratio
Aluminum alloy	70	250	0.33
Titanium alloy	105	850	0.36
Steel alloy	205	550	0.27
Magnesium alloy	45	170	0.35

2.23 A cylindrical rod 500 mm (20.0 in) long, having a diameter of 12.7 mm (0.50 in), is to be subjected to a tensile load. If the rod is to experience neither plastic deformation nor an elongation of more than 1.3 mm (0.05 in) when the applied load is 29000 N (6500 lbf), which of the four metals or alloys listed below are possible candidates? Justify your choice(s).

Material	Modulus of Elasticity/GPa	Yield Strength/MPa	Tensile Strength/MPa
Aluminum alloy	70	255	420
Brall alloy	100	345	420
Copper	110	210	275
Steel alloy	207	450	550

2.24 Figure 2.20 shows the tensile engineering stress-strain behavior for a steel alloy.

(1) What is the modulus of elasticity?

(2) What is the proportional limit?

(3) What is the yield strength at a strain offset of 0.002?

(4) What is the tensile strength?

2.25 A cylindrical specimen of a brass alloy having a length of 100 mm (4 in) must elongate only 5 mm (0.2 in) when a tensile load of 100000 N (22500 lbf) is applied. Under these circumstances what must be the radius of the specimen? Consider this brass alloy to have the stress-strain behavior shown in Figure 2.12.

2.26 A load of 140000 N (31500 lbf) is applied to a cylindrical specimen of a steel alloy (displaying the stress-strain behavior shown in Figure 2.20) that has a cross-sectional diameter of 10 mm (0.40 in).

(1) Will the specimen experience elastic and/ or plastic deformation? Why?

(2) If the original specimen length is 500 mm (20 in), how much will it increase in length when this load is applied?

2.27 A bar of a steel alloy that exhibits the stress-strain behavior shown in Figure 2.20 is subjected to a tensile load; the specimen is 375 mm (14.8 in) long and of square cross section 5.5 mm (0.22 in) on a side.

(1) Compute the magnitude of the load necessary to produce an elongation of 2.25 mm (0.088 in).

(2) What will be the deformation after the load has been released?

2.28 Show that Equations 2.18a and 2.18b are valid when there is no volume change during deformation.

2.29 Demonstrate that Equation 2.16, the expression defining true strain, may also be represented by

$$\varepsilon_T = \ln\left(\frac{A_0}{A_i}\right)$$

when specimen volume remains constant during deformation. Which of these two expressions is more valid during necking? Why?

2.30 The following true stresses produce the corresponding true plastic strains for a brass alloy: What true stress is necessary to produce a true plastic strain of 0.21?

True Stress/psi	True Strain
60000	0.15
70000	0.25

2.31 For a brass alloy, the following engineering stresses produce the corresponding plastic engineering strains, prior to necking:

Engineering Stress/MPa	Engineering Strain
315	0.105
340	0.220

On the basis of this information, compute the *engineering* stress necessary to produce an engineering strain of 0.28.

Chapter 2

IMPORTANT TERMS AND CONCEPTS

Uniaxial Tensile Test	单向拉伸试验	Plastic Deformation	塑性变形
Tensile Stress	拉应力	Fracture	断裂
Compressive Stress	压应力	Stiffness	刚度
Torsion	扭转	Modulus of Elasticity	弹性模量
Elasticity	弹性	Young'S Modulus	杨氏模量
Engineering Stress	工程应力	Shear Modulus	剪切模量
Engineering Strain	工程应变	Isotropic Material	各向同性材料
Anelasticity/ Viscoelasticity	滞弹性	Anisotropic Material	各向异性材料
Resilience	弹性变形功	Lüders	吕德斯
Modulus of Resilience	弹性比功/弹性比能/应变比能	Yield Point Phenomenon	屈服点现象
Hooke'S Law	胡克定律	Yielding	屈服
Proportional Limit	比例极限	Strength	强度
Elastic Deformation	弹性变形	Yield Strength	屈服强度

Tensile Strength	抗拉强度	True Stress	真应力
Ductility/Plasticity	塑性	True Strain	真应变
Percent Elongation(El)	伸长率	Strain-Hardening Exponent	应变硬化指数
Percent Reduction in Area(Ra)	断面收缩率	Hardness	硬度
Poisson'S Ratio	泊松比	Indentation	压痕
Shear Stress	剪切应力	Brinell Hardness	布氏硬度
Shear Strain	剪切应变	Rockwell Hardness	洛氏硬度
Ductile Material	韧性材料	Vickers Hardness	维氏硬度
Brittle Material	脆性材料	Knoop Hardness	努氏硬度
Impact Toughness	冲击韧度	Martensitic Phase Transformation	马氏体相变

Chapter 3

Micro-fracture of metals

Learning Objective

After studying this chapter you should be able to do the following:

1. Describe the mechanism of crack nucleation of fracture.
2. Understanding the process or both ductile and brittle modes of fracture.

3.1 Introduction

The separation or fragmentation of a solid body into two or more parts, under the action of stresses, is called fracture. The subject of fracture is vast and involves disciplines as diverse as solid-state physics, materials science, and continuum mechanics. Fracture of a material by cracking can occur in many ways, principally the following:

(1) Slow application of external loads.

(2) Rapid application of external loads (impact).

(3) Cyclic or repeated loading (fatigue).

(4) Time-dependent deformation (creep).

(5) Internal stresses, such as thermal stresses caused by anistropy of the thermal expansion coefficient or temperature differences in a body.

(6) Environmental effects (stress corrosion cracking, hydrogen embrittlement, liquid metal embrittlement, etc.)

Figure 3.1 shows, schematically, some important fracture modes in metals. These different modes will be analyzed in some detail in this chapter. Metals fail by two broad classes of mechanisms: ductile and brittle failure.

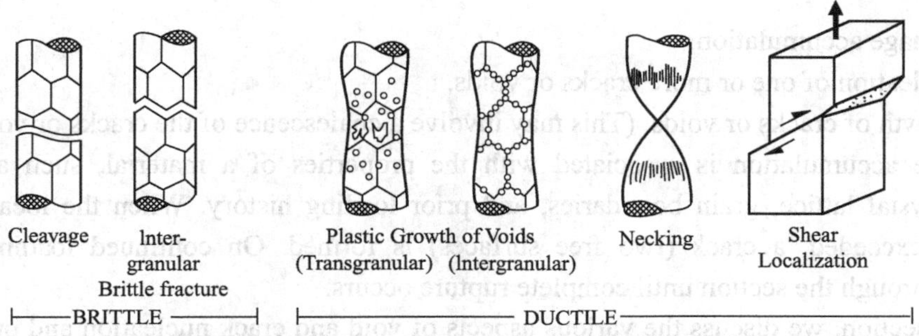

Figure 3.1 Some important fracture modes in metals.

3.1.1 Ductile fracture

Ductile failure occurs by:

(1) the nucleation, growth, and coalescence of voids,

(2) continuous reduction in the metal's cross-sectional area until it is equal to zero,

(3) shearing along a plane of maximum shear.

Ductile failure by void nucleation and growth usually starts at second-phase particles. If these particles are spread throughout the interiors of the grains, the fracture will be transgranular (or transcrystalline). If these voids are located preferentially at grain boundaries, fracture will occur in an intergranular (or intercrystalline) mode. The appearance of a ductile fracture, at high magnification (500× or higher) is of a surface with indentations, as if marked by an ice cream scooper. This surface morphology is appropriately called dimpled. Rupture by total necking is very rare, because most metals contain second-phase particles that act as initiation sites for voids. However, high-purity metals, such as copper, nickel, gold, and other very ductile materials, fail with very high reductions in their areas.

3.1.2 Brittle fracture

Brittle fracture is characterized by the propagation of one or more cracks through the structure. While totally elastic fracture describes the behavior of most ceramics fairly well, metals and some polymers undergo irreversible deformation at the tip of the crack, which affects its propagation. For metals and ceramics, two modes of crack propagation: transgranular fracture (or cleavage) and intergranular fracture are observed. For energy-related reasons, a crack will tend to take the path of least resistance. If this path lies along the grain boundaries, the fracture will be intergranular. Often, a crack also tends to run along specific crystallographic planes, as is the case for brittle fracture in steel. Upon observation at high magnification, transgranular brittle fracture is characterized by clear, smooth facets that have the size of the grains. In steel, brittle fracture has the typical shiny appearance, while ductile fracture has a dull, grayish aspect. In addition to brittle fracture, polymers undergo a mode of fracture called crazing, in which the polymer chains ahead of a crack align themselves along the tensile axis, so that the stress concentration is released.

3.2 Process of fracture

Metals are characterized by a highly mobile dislocation density, and they generally show a ductile fracture. In most cases, the process of fracture can be subdivided into the following categories:

(1) Damage accumulation.

(2) Nucleation of one or more cracks or voids.

(3) Growth of cracks or voids. (This may involve a coalescence of the cracks or voids.)

Damage accumulation is associated with the properties of a material, such as its atomic structure, crystal lattice, grain boundaries, and prior loading history. When the local strength or ductility is exceeded, a crack (two free surfaces) is formed. On continued loading, the crack propagates through the section until complete rupture occurs.

In this section, we discuss the various aspects of void and crack nucleation and propagation in metals.

3.2.1 Crack Nucleation

Nucleation of a crack in a perfect crystal essentially involves the rupture of interatomic bonds.

The stress necessary to do this is the theoretical cohesive stress, starting from an expression for interatomic forces. From this expression, we see that ordinary materials break at much lower stresses than do perfect crystals (on the order of $E/10^4$), where E is Young's modulus of the material. The explanation of this behavior lies in the existence of surface and internal defects that act as preexisting cracks and in the plastic deformation that precedes fracture. When both plastic deformation and fracture are eliminated, for example, stresses (in ''whiskers'') on the order of the theoretical cohesive stresses are obtained.

Crack nucleation mechanisms vary according to the type of material: brittle, semibrittle, or ductile. The brittleness of a material has to do with the behavior of dislocations in the region of crack nucleation. In highly brittle materials the dislocations are practically immobile, in semibrittle materials dislocations are mobile, but only on a restricted number of slip planes, and in ductile materials there are no restrictions on the movement of dislocations other than those inherent in the crystalline structure of the material.

The exposed surface of a brittle material can suffer damage by mechanical contact with even microscopic dust particles. If a glass fiber without surface treatment were rolled over a tabletop, it would be seriously damaged mechanically.

Any heterogeneity in a material that produces a stress concentration can nucleate cracks. For example, steps, striations, depressions, holes, and so on act as stress raisers on apparently perfect surfaces. In the interior of the material, there can exist voids, air bubbles, second-phase particles, etc. Crack nucleation will occur at the weakest of these defects, where the conditions would be most favorable. We generally assume that the sizes as well as the locations of defects are distributed in the material according to some function of standard distribution whose parameters are adjusted to conform to experimental data. In this assumption, there is no explicit consideration of the nature or origin of the defects.

In semibrittle materials, there is a tendency for slip initially, followed by fracture on well-defined crystallographic planes. That is, there exists a certain inflexibility in the deformation process, and the material, not being able to accommodate localized plastic strains, initiates a crack to relax stresses.

Various models are based on the idea of crack nucleation at an obstruction site. For example, the intersection of a slip band with a grain boundary, another slip band, and so on, would be an obstruction site.

In crystalline solids, cracks can be nucleated by the grouping of dislocations piled up against a barrier. Such cracks are called Zener-Stroh cracks.1 High stresses at the head of a pileup are relaxed by crack nucleation, as shown in Figure 3.2, but this would occur only in the case where there is no relaxation of stresses by the movement of dislocations on the other side of the barrier. Depending on the slip geometry in the two parts and the kinetics of the motion and multiplication of dislocations, such a combination of events could occur. Figure 3.3(a) shows a bicrystal that has a slip band in grain I. The stress concentration at

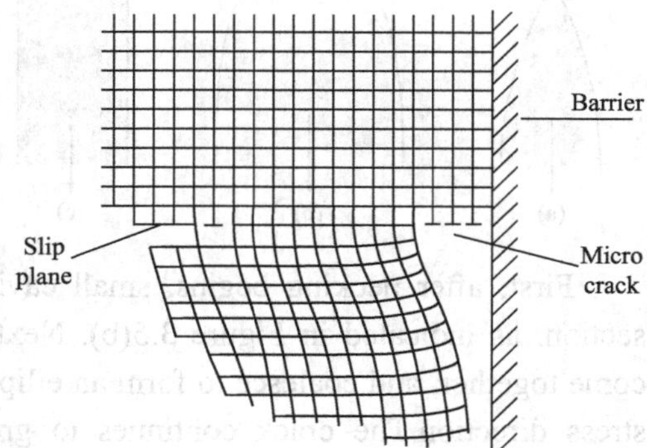

Figure 3.2 Grouping of dislocations piled up at a barrier and leading to the formation of a microcrack (Zener-Stroh crack).

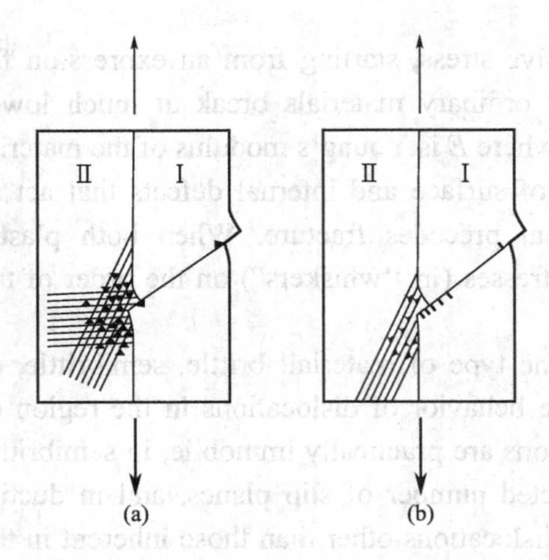

Figure 3.3 Bicrystal with a slip band in grain I. (a) The stress concentration at the boundary of the barrier due to slip band is fully relaxed by multiple slip. (b) The stress concentration is only partially relaxed, resulting in a crack at the boundary.

the barrier due to the slip band is completely relaxed by slip on two systems in grain II. Figure 3.3(b) shows the case of only a partial relaxation and the resulting appearance of a crack at the barrier. Lattice rotation associated with the bend planes and deformation twins can also nucleate cracks.

3.2.2 Ductile Fracture

Ductile fracture surfaces will have their own distinctive features on both macroscopic and microscopic levels. Figure 3.4 shows schematic representations for two characteristic macroscopic fracture profiles. The configuration shown in Figure 3.4(a) is found for extremely soft metals, such as pure gold and lead at room temperature, and other metals, polymers, and inorganic glasses at elevated temperatures. These highly ductile materials neck down to a point fracture, showing virtually 100% reduction in area. The most common type of tensile fracture profile for ductile metals is that represented in Figure 3.4(b), where fracture is preceded by only a moderate amount of necking. The fracture process normally occurs in several stages (Figure 3.2).

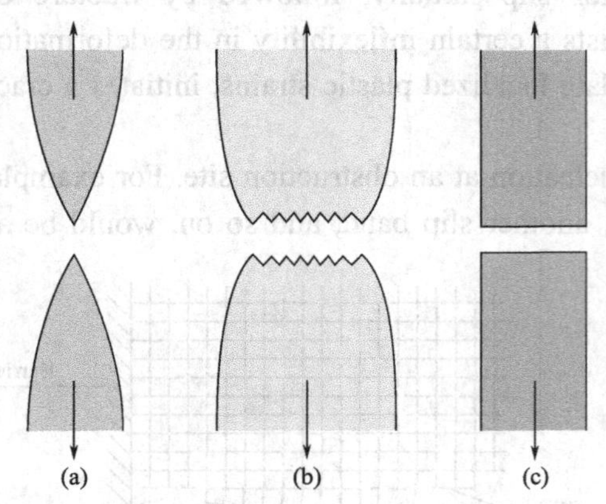

Figure 3.4 (a) Highly ductile fracture in which the specimen necks down to a point. (b) Moderately ductile fracture after some necking. (c) Brittle fracture without any plastic deformation.

First, after necking begins, small cavities, or microvoids, form in the interior of the cross section, as indicated in Figure 3.5(b). Next, as deformation continues, these microvoids enlarge, come together, and coalesce to form an elliptical crack, which has its long axis perpendicular to the stress direction.The crack continues to grow in a direction parallel to its major axis by this microvoid coalescence process [Figure 3.5(c)]. Finally, fracture ensues by the rapid propagation of a crack around the outer perimeter of the neck [Figure 3.5(d)], by shear deformation at an angle of about 45°with the tensile axis—this is the angle at which the shear stress is a maximum. Sometimes

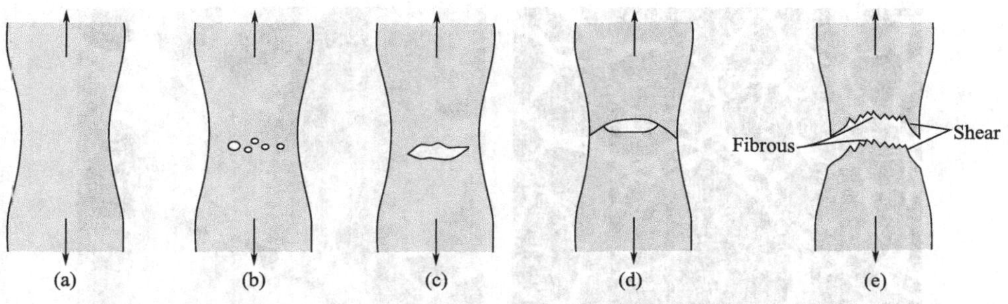

Figure 3.5 Stages in the cup-and-cone fracture. (a) Initial necking. (b) Smallcavity formation. (c) Coalescence of cavities to form a crack. (d) Crack propagation. (e) Final shear fracture at a 45° angle relative to the tensile direction. (From K. M. Ralls, T. H. Courtney, and J.Wulff, Introduction to Materials Science and Engineering, 468.)

a fracture having this characteristic surface contour is termed a cup-and-cone fracture because one of the mating surfaces is in the form of a cup, the other like a cone. In this type of fractured specimen [Figure 3.6(a)], the central interior region of the surface has an irregular and fibrous appearance, which is indicative of plastic deformation.

Figure 3.6 (a) Cup-and-cone fracture in aluminum. (b) Brittle fracture in a mild steel.

Much more detailed information regarding the mechanism of fracture is available from microscopic examination, normally using scanning electron microscopy. Studies of this type are termed fractographic. The scanning electron microscope is preferred for fractographic examinations since it has a much better resolution and depth of field than does the optical microscope; these characteristics are necessary to reveal the topographical features of fracture surfaces.

When the fibrous central region of a cup-and-cone fracture surface is examined with the electron microscope at a high magnification, it will be found to consist of numerous spherical "dimples" [Figure 3.7(a)]; this structure is characteristic of fracture resulting from uniaxial tensile failure. Each dimple is one half of a microvoid that formed and then separated during the fracture process. Dimples also form on the shear lip of the cup-and-cone fracture. However, these will be elongated or C-shaped, as shown in Figure 3.7(b). This parabolic shape may be indicative of shear failure. Furthermore, other microscopic fracture surface features are also possible. Fractographs such as those shown in Figures 3.7(a) and 3.7(b) provide valuable information in the analyses of fracture, such as the fracture mode, the stress state, and the site of crack initiation.

In ductile materials, the role of plastic deformation is very important. The important feature is the flexibility of slip. Dislocations can move on a large number of slip systems and even cross from one plane to another (in cross-slip). Consider the deformation of a single crystal of copper, a ductile metal, under uniaxial tension. The single crystal undergoes slip throughout its section. There is no

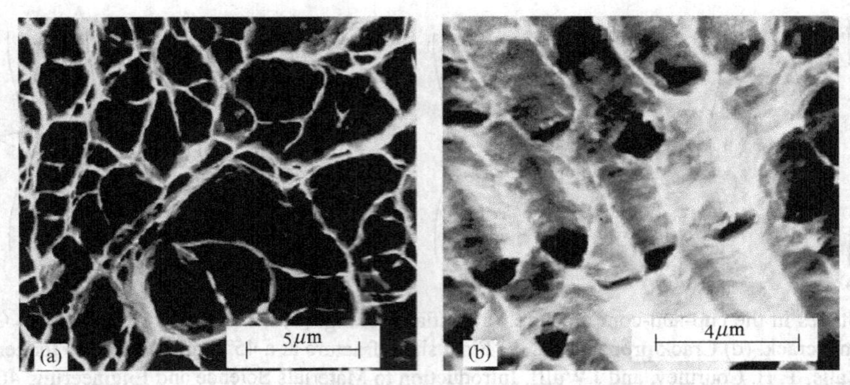

Figure 3.7 (a) Scanning electron fractograph showing spherical dimples characteristic of ductile fracture resulting from uniaxial tensile loads. 3300×. (b) Scanning electron fractograph showing parabolic-shaped dimples characteristic of ductile fracture resulting from shear loading. 5000×. (From R.W. Hertzberg, Deformation and Fracture Mechanics of Engineering Materials, 3rd edition.)

nucleation of cracks, and the crystal deforms plastically until the start of plastic instability, called necking. From this point onward, the deformation is concentrated in the region of plastic instability until the crystal separates along a line or a point [Figure 3.8(a)]. In the case of a cylindrical sample, a soft single crystal of a metal such as copper will reduce to a point fracture. Figure 3.8(b) shows an example of such a fracture in a single crystal of copper. However, if, in a ductile material, there are microstructural elements such as particles of a second phase, internal interfaces, and so on, then microcavities may be nucleated in regions of high stress concentration in a manner similar to that of semibrittle materials, except that, due to the ductile material's large plasticity, cracks generally do not propagate from these cavities. The regions between the cavities, though, behave as small test samples that elongate and break by plastic instability, as described for the single crystal.

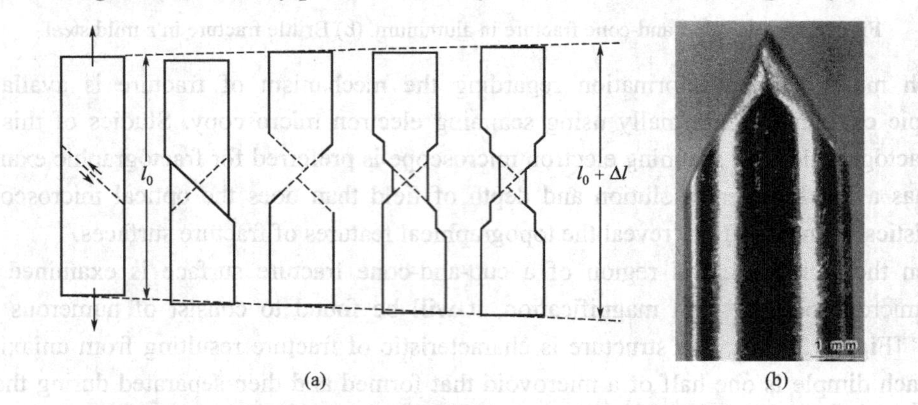

Figure 3.8 (a) Failure by shear (glide) in a pure metal. (b) A point fracture in a soft single-crystal sample of copper.

If the second-phase particles are brittle and the matrix is ductile, the former will not be able to accommodate the large plastic strains of the matrix, and consequently, these brittle particles will break in the very beginning of plastic deformation. In case the particle/matrix interface is very weak, interfacial separation will occur. In both cases, microcavities are nucleated at these sites (Figure 3.9). Generally, the voids nucleate after a few percent of plastic deformation, while the final separation may occur around 25%. The microcavities grow with slip, and the material between the cavities can be visualized as a small tensile test piece. The material between the voids undergoes necking on a microscopic scale, and the voids join together. However, these microscopic necks do not contribute significantly to the total elongation of the material. This mechanism of initiation, growth, and coalescence of microcavities gives the fracture surface a characteristic appearance. When viewed in the scanning electron microscope, such a fracture appears to consist of small dimples, which represent the microcavities after coalescence. In many of these dimples, one can see the inclusions that were responsible for the void nucleation (Figure 3.10). At times, due to unequal triaxial stresses, these voids are elongated in one or the other direction. We describe the process of fracture by void nucleation, growth, and coalescence in some detail because of its great importance in metals.

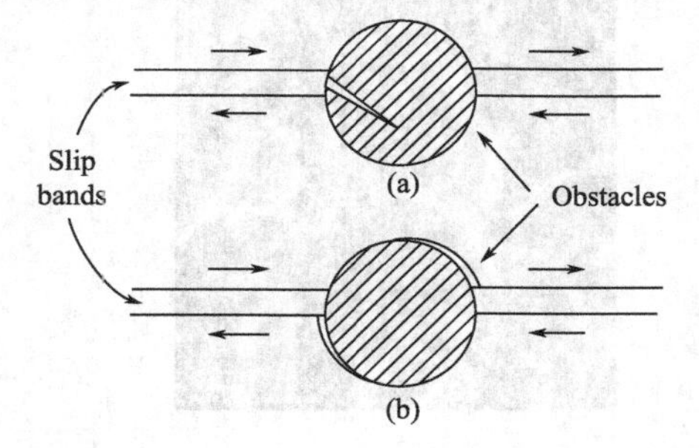

Figure 3.9 Nucleation of a cavity at a second-phase particle in a ductile material.

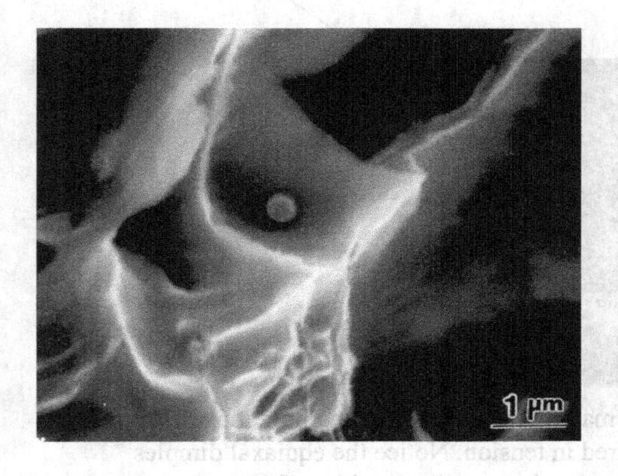

Figure 3.10 Scanning electron micrograph of dimple fracture resulting from the nucleation, growth, and coalescence of microcavities. The micrograph shows an inclusion, which served as the microcavity nucleation sites.

Fracture by Void Nucleation, Growth, and Coalescence

Figure 3.11 shows the classic cup-and-cone fracture observed in many tensile specimens with a cylindrical cross section. The configuration is typical of ductile fracture, and upon observation at a higher magnification (1000× or higher, best done in a scanning electron microscope), one sees the typical ''dimple" features. The dimples are equiaxal in the central portion of the fracture and tend to be inclined in the sidewalls of the ''cup." The top two pictures show scanning electron micrographs of these two areas. In the central region fracture is essentially tensile, with the surface perpendicular to the tensile axis.

On the sides, the fracture has a strong shear character, and the dimples show the typical "inclined" morphology, i.e., they appear to be elliptical with one side missing. Figure 3.12 shows, in a very schematic fashion, what is thought to occur in the specimen that leads to failure. Voids nucleate and grow in the interior of the specimen when the overall plastic strain reaches a critical level. The voids grow until they coalesce. Initially equiaxial, their shape changes in accordance with the overall stress field. As the voids coalesce, they expand into adjoining areas, due to the stress concentration effect. When the center of the specimen is essentially separated, this failure will grow toward the outside. Since the elastic and plastic constraints change, the plane of maximum shear (approximately 45° to the tensile axis) is favored, and further growth will take place along these planes, which form the sides of the cup. Although it is easy to describe this process in a qualitative way, an analytical derivation is very complex and involves plasticity theory, which is beyond the scope of the text. Figure 3.13 shows the sequence of ductile fracture propagation, with the formation of dimples. The dimples are produced by voids nucleating ahead of the principal crack

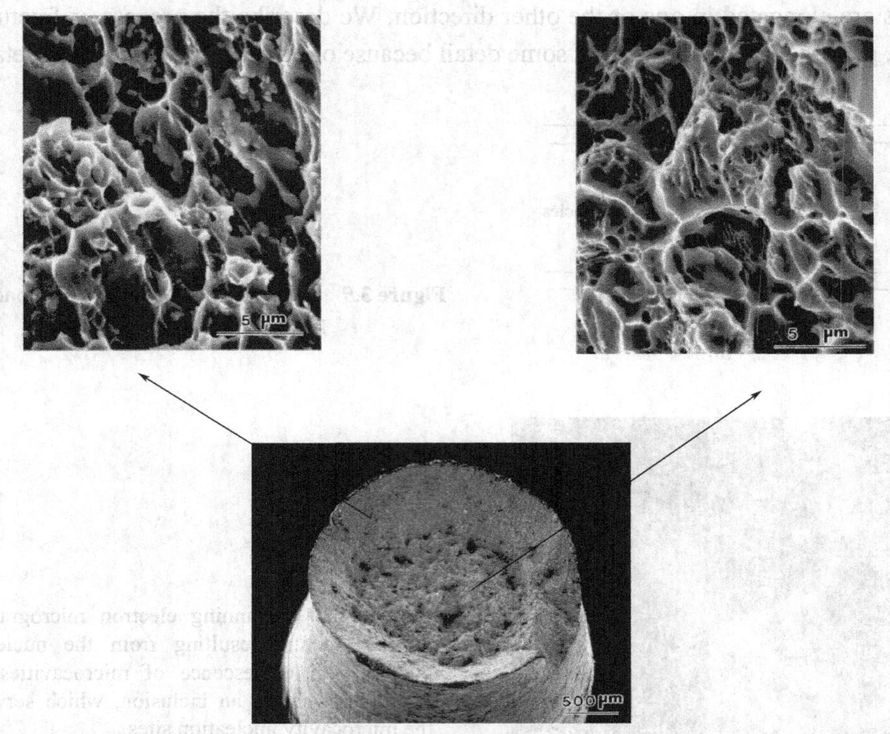

Figure 3.11 Scanning electron micrographs at low magnification (center) and high magnification (right and left) of AISI 1008 steel specimen ruptured in tension. Notice the equiaxal dimples in the central region and elongated dimples on the shear walls, the sides of the cup.

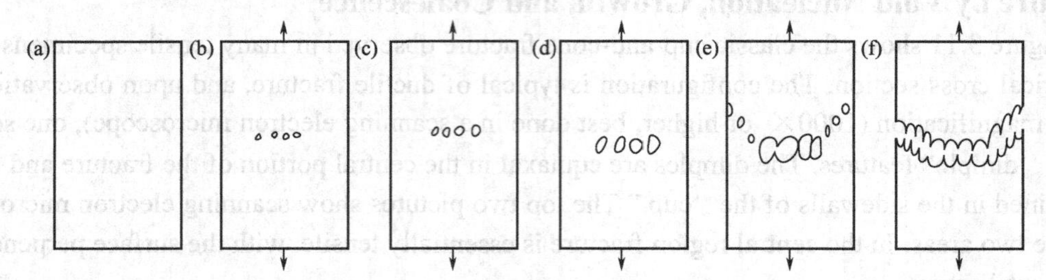

Figure 3.12 Schematic sequence of events leading to the formation of a cup-and-cone fracture.

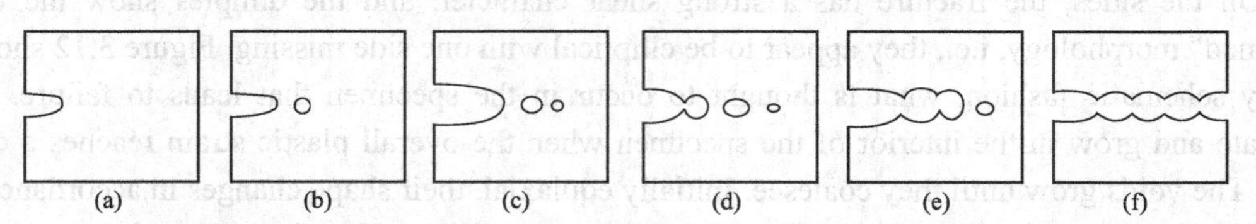

Figure 3.13 Sequence of events in the propagation of ductile fracture by nucleation, growth, and coalescence of voids.

[Figures 3.13(a) and (b)], which has a blunted tip because of the plasticity of the material. The void ahead of the crack grows [Figure 3.13(c)] and eventually coalesces with the main crack [Figure 3.13(d)]. New voids nucleate ahead of the growing crack, and the process repeats itself.

3.2.3 Brittle fracture

Brittle fracture takes place without any appreciable deformation, and by rapid crack propagation. The direction of crack motion is very nearly perpendicular to the direction of the applied tensile stress and yields a relatively flat fracture surface, as indicated in Figure 3.1.

Fracture surfaces of materials that failed in a brittle manner will have their own distinctive patterns; any signs of gross plastic deformation will be absent. For example, in some steel pieces, a series of V-shaped "chevron" markings may form near the center of the fracture cross section that point back toward the crack initiation site [Figure 3.14(a)]. Other brittle fracture surfaces contain lines or ridges that radiate from the origin of the crack in a fanlike pattern [Figure 3.14(b)]. Often, both of these marking patterns will be sufficiently coarse to be discerned with the naked eye. For very hard and fine-grained metals, there will be no discernible fracture pattern. Brittle fracture in amorphous materials, such as ceramic glasses, yields a relatively shiny and smooth surface.

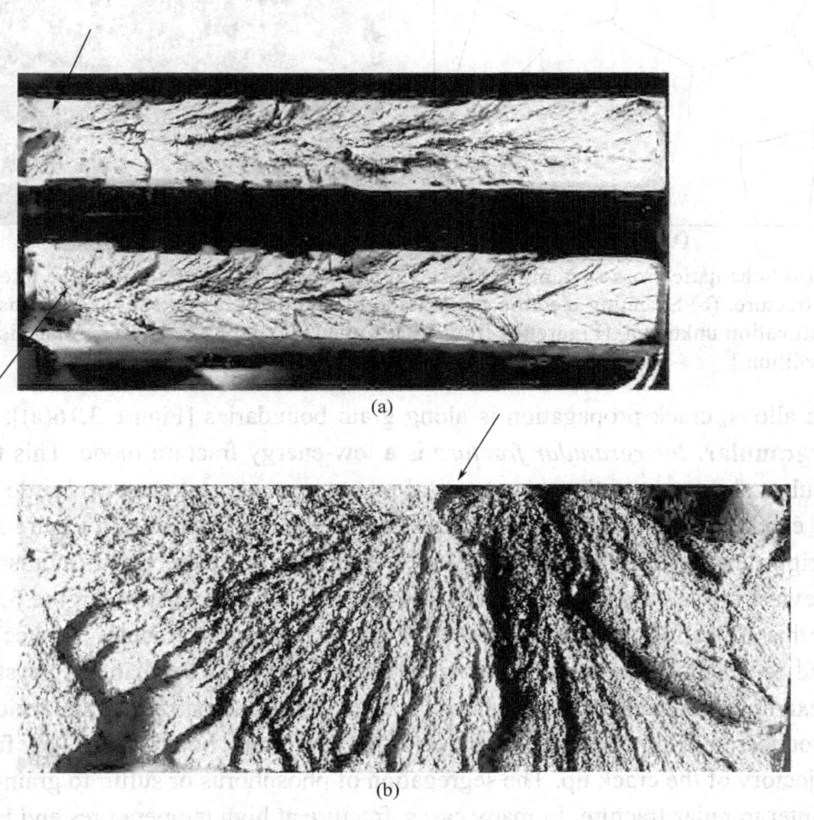

Figure 3.14 (a) Photograph showing V-shaped "chevron" markings characteristic of brittle fracture. Arrows indicate origin of crack. Approximately actual size. (b) Photograph of a brittle fracture surface showing radial fan-shaped ridges. Arrow indicates origin of crack. Approximately 2×. [(a) From R.W. Hertzberg, *Deformation and Fracture Mechanics of Engineering Materials,* 3rd edition. (b) Reproduced with permission from D. J.Wulpi, *Understanding How Components Fail,* American Society for Metals, Materials Park, OH, 1985.]

For most brittle crystalline materials, crack propagation corresponds to the successive and repeated breaking of atomic bonds along specific crystallographic planes [Figure 3.15(a)]; such a process is termed cleavage. This type of fracture is said to be **transgranular** (or transcrystalline), because the fracture cracks pass through the grains. Macroscopically, the fracture surface may have a grainy or faceted texture, as a result of changes in orientation of the cleavage planes from grain to grain. This cleavage feature is shown at a higher magnification in the scanning electron micrograph of Figure 3.15(b).

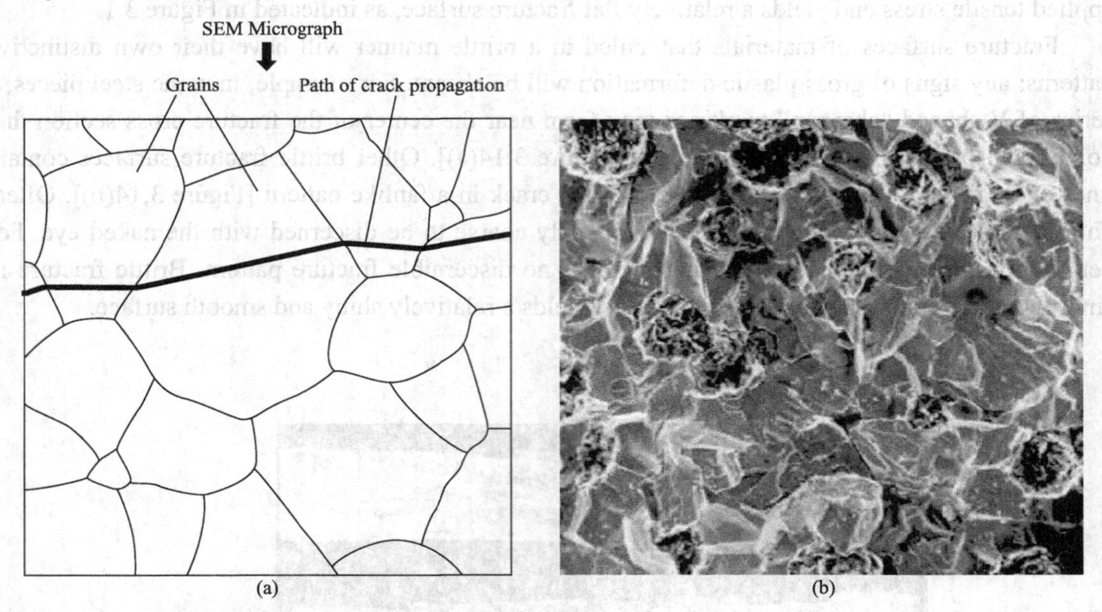

Figure 3.15 (a) Schematic cross-section profile showing crack propagation through the interior of grains for transgranular fracture. (b) Scanning electron fractograph of ductile cast iron showing a transgranular fracture surface. Magnification unknown. [Figure (b) from V. J. Colangelo and F. A. Heiser, Analysis of Metallurgical Failures, 2nd edition.]

In some alloys, crack propagation is along grain boundaries [Figure 3.16(a)]; this fracture is termed **intergranular.** *Intergranular fracture* is a low-energy fracture mode. This type of fracture normally results subsequent to the occurrence of processes that weaken or embrittle grain boundary regions. The crack follows the grain boundaries, as shown schematically in Figure 3.16, giving the fracture a bright and reflective appearance on a macroscopic scale. On a microscopic scale, the crack may deviate around a particle and make some microcavities locally. Figure 3.16(b) shows an example of this deviation in a micrograph of an intergranular fracture in steel. Intergranular fractures tend to occur when the grain boundaries are more brittle than the crystal lattice. This occurs, for example, in stainless steel when it is accidentally sensitized. This accident in the heat treatment produces a film of brittle carbides along the grain boundaries. The film is then the preferred trajectory of the crack tip. The segregation of phosphorus or sulfur to grain boundaries can also lead to intergranular fracture. In many cases, fracture at high temperatures and in creep tends to be intergranular.

The most brittle form of fracture is cleavage fracture. The tendency for a cleavage fracture increases with an increase in the strain rate or a decrease in the test temperature of a material. This is shown, typically, by a ductile—brittle transition in steel in a Charpy impact test (Figure 3.17).

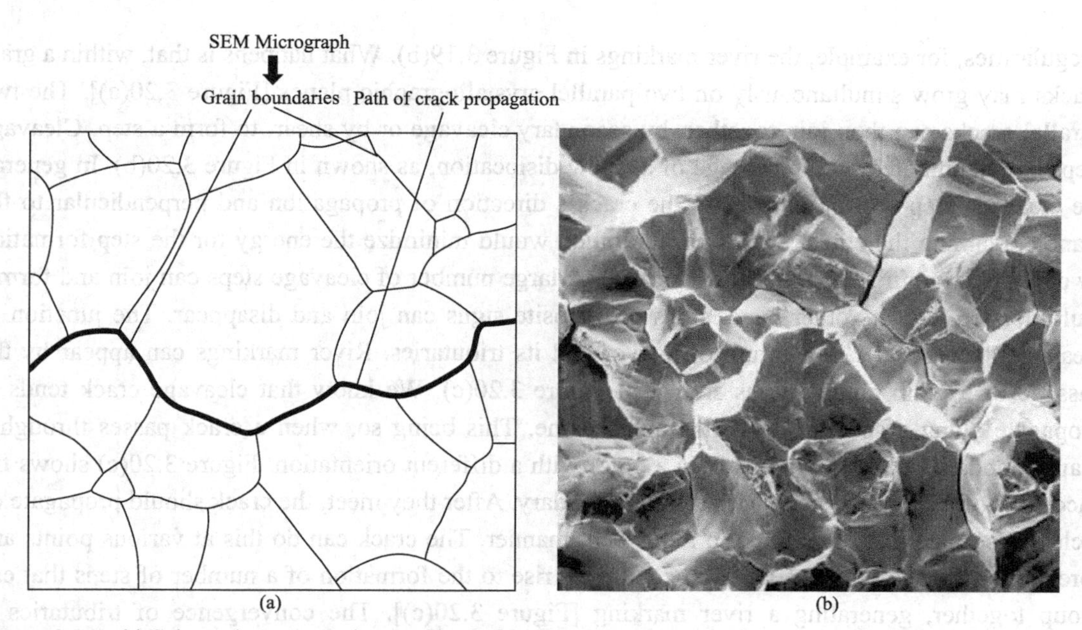

Figure 3.16 (a) Schematic cross-section profile showing crack propagation along grain boundaries for intergranular fracture. (b) Scanning electron fractograph showing an intergranular fracture surface. 50×. [Figure (b) reproduced with permission from ASM Handbook, Vol. 12, Fractography, ASM International, Materials Park, OH, 1987.]

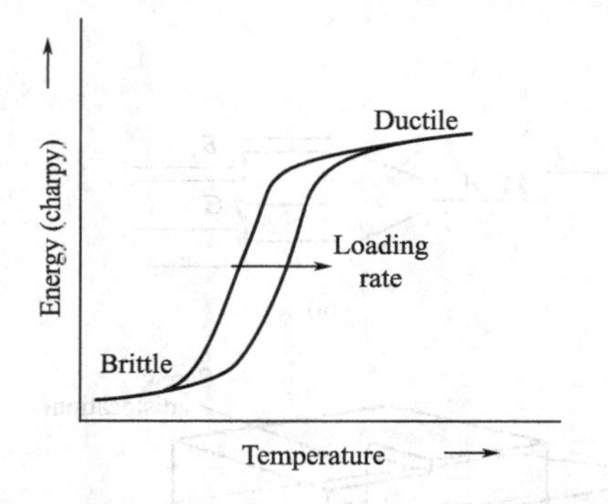

Figure 3.17 Schematic of ductile–brittle transition in steel and the effect of loading rate.

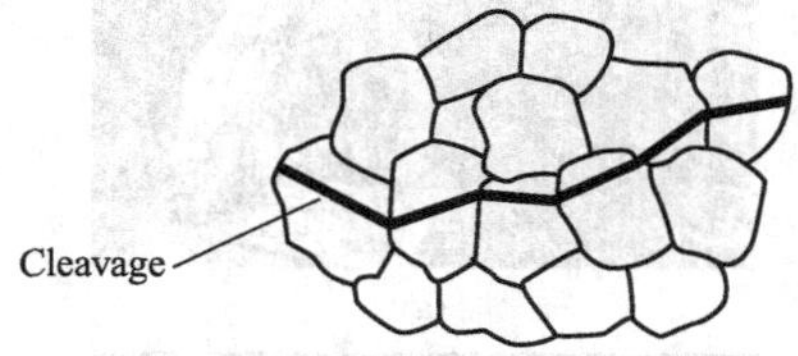

Cleavage

Figure 3.18 Propagation of transgranular cleavage.

The ductile—brittle transition temperature (DBTT) increases with an increase in the strain rate. Above the DBTT the steel shows a ductile fracture, while below the DBTT it shows a brittle fracture. The ductile fracture needs a lot more energy than the brittle fracture. Cleavage occurs by direct separation along specific crystallographic planes by means of a simple rupturing of atomic bonds [Figure 3.18]. Iron, for example, undergoes cleavage along its cubic planes (100). This gives the characteristic flat surface appearance within a grain on the fracture surface. There is evidence that some kind of plastic yielding and dislocation interaction is responsible for cleavage fracture.

Earlier, we mentioned that cleavage occurs along specific crystallographic planes. As in a polycrystalline material, the adjacent grains have different orientations; the cleavage crack changes direction at the grain boundary in order to continue along the given crystallographic planes. The cleavage facets seen through the grains have a high reflectivity, which gives the fracture surface a shiny appearance [Figure 3.19(a)]. Sometimes the cleavage fracture surface shows some small irregularities, for example, the river markings in Figure 3.19(b). What happens is that, within a grain, cracks may grow simultaneously on two parallel crystallographic planes [Figure 3.20(a)]. The two parallel cracks can then join together, by secondary cleavage or by shear, to form a step. Cleavage steps can be initiated by the passage of a screw dislocation, as shown in Figure 3.20(b). In general, the cleavage step will be parallel to the crack's direction of propagation and perpendicular to the plane containing the crack, as this configuration would minimize the energy for the step formation by creating a minimum of additional surface. A large number of cleavage steps can join and form a multiple step. On the other hand, steps of opposite signs can join and disappear. The junction of cleavage steps results in a figure of a river and its tributaries. River markings can appear by the passage of a grain boundary, as shown in Figure 3.20(c). We know that cleavage crack tends to propagate along a specific crystallographic plane. This being so, when a crack passes through a grain boundary, it has to propagate in a grain with a different orientation. Figure 3.20(c) shows the encounter of a cleavage crack with a grain boundary. After they meet, the crack should propagate on a cleavage plane that is oriented in a different manner. The crack can do this at various points and spread into the new grain. Such a process gives rise to the formation of a number of steps that can group together, generating a river marking [Figure 3.20(c)]. The convergence of tributaries is always in the direction of flow of the river (i.e., "downstream"). This fact furnishes the possibility of determining the local direction of propagation of crack in a micrograph.

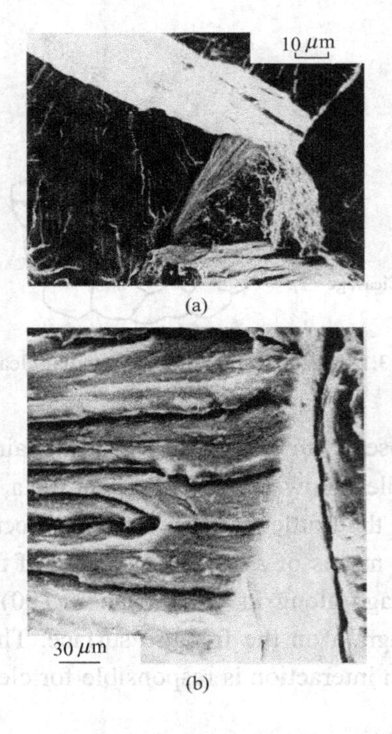

Figure 3.19 (a) Cleavage facets in 300-M steel (scanning electron micrograph). (b) River markings on a cleavage facet in 300-M steel.

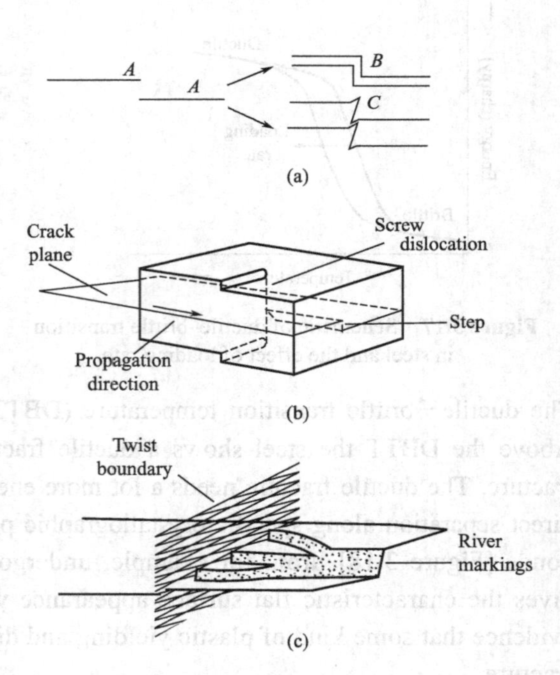

Figure 3.20 Formation of cleavage steps. (a) Parallel cracks (*A*, *A*) join together by cleavage (*B*) or shear (*C*). (b) Cleavage step initiation by the passage of a screw dislocation. (c) Formation of river markings after the passage of a grain boundary.

Under normal circumstances, face-centered cubic (FCC) metals do not show cleavage. In these metals, a large amount of plastic deformation will occur before the stress necessary for cleavage is reached. Cleavage is common in body-centered cubic (BCC) and hexagonal close-packed (HCP) structures, particularly in iron and low carbon steels (BCC). Tungsten, molybdenum, and chromium (all BCC) and zinc, beryllium, and magnesium (all HCP) are other examples of metals that commonly show cleavage.

Quasi cleavage is a type of fracture that is formed when cleavage occurs on a very fine scale and on cleavage planes that are not very well defined. Typically, one sees this type of fracture in quenched and tempered steels. These steels contain tempered martensite and a network of carbide particles whose size and distribution can lead to a poor definition of cleavage planes in the austenite grain. Thus, the real cleavage planes are exchanged for small and ill-defined cleavage facets that initiate at the carbide particles. Such small facets can give the appearance of a much more ductile fracture than that of normal cleavage, and generally, river markings are not observed.

SUMMARY

Ductile Fracture

Fracture, in response to tensile loading and at relatively low temperatures, may occur by ductile and brittle modes, both of which involve the formation and propagation of cracks. For ductile fracture, evidence will exist of gross plastic deformation at the fracture surface. In tension, highly ductile metals will neck down to essentially a point fracture; cup-and-cone mating fracture surfaces result for moderate ductility. Cracks in ductile materials are said to be stable (i.e., resist extension without an increase in applied stress); and inasmuch as fracture is noncatastrophic, this fracture mode is almost always preferred.

Brittle Fracture

For brittle fracture, cracks are unstable, and the fracture surface is relatively flat and perpendicular to the direction of the applied tensile load. Chevron and ridgelike patterns are possible, which indicate the direction of crack propagation. Transgranular (through-grain) and intergranular (between-grain) fractures are found in brittle polycrystalline materials.

IMPORTANT TERMS AND CONCEPTS

Ductile Fracture
Brittle Fracture
Cleavage
Transgranular Fracture
Intergranular Fracture

REFERENCES

1. C. Zener, The Fracturing of Metals (Metals Park, OH: ASM, 1948)
2. M. F. Ashby, Materials Selection in Mechanical Design, 2nd ed. Elsevier, 1999
3. Colangelo, V. J. and F. A. Heiser, Analysis of Metallurgical Failures, 2nd edition, Wiley, New York, 1987
4. Collins, J.A., Failure of Materials in Mechanical Design, 2nd edition,Wiley, New York, 1993
5. Courtney, T. H., Mechanical Behavior of Materials, 2nd edition, McGraw-Hill, New York, 2000
6. Dieter, G. E., Mechanical Metallurgy, 3rd edition, McGraw-Hill, New York, 1986
7. Esaklul, K. A., Handbook of Case Histories in Failure Analysis, ASM International, Materials Park, OH, 1992 and 1993. In two volumes
8. Hertzberg, R. W., Deformation and Fracture Mechanics of Engineering Materials, 4th edition, Wiley, New York, 1996

9. Stevens, R. I., A. Fatemi, R. R. Stevens, and H. O. Fuchs, Metal Fatigue in Engineering, 2nd edition, Wiley, New York, 2000
10. Tetelman, A. S. and A. J. McEvily, Fracture of Structural Materials, Wiley, New York, 1967. Reprinted by Books on Demand, Ann Arbor, MI
11. Wulpi, D. J., Understanding How Components Fail, 2nd edition, ASM International, Materials Park, OH, 1999

QUESTIONS AND PROBLEMS

3.1 In Figure 3.6, mechanical twinning has generated microcracks that, in subsequent tensile tests, weakened the specimen. The ultimate tensile strength of tungsten is 1.2 GPa, and its fracture toughness is approximately 70 MPa $m^{1/2}$. By how much is the fracture stress decreased due to the presence of the microcracks?

3.2 Explain why FCC metals show a ductile fracture even at low temperatures, while BCC metals do not.

3.3 Show, by a sequence of sketches, how the neck in pure copper and in copper with 15% volume fraction of a second phase will develop. Using values from Figure 3.9, show the approximate configuration of the final neck.

3.4 Alumina specimens contain flaws introduced during processing; these flaws are, approximately, the grain size. Plot the fracture stress vs. grain size (for grains below 200 μm), knowing that the fracture toughness for alumina is equal to 4 MPa $m^{1/2}$. Assume $Y = 1$.

Chapter 3

IMPORTANT TERMS AND CONCEPTS

Ductile Fracture	韧性断裂	"River" Markings	河流花样
Microcrack Nucleation	微裂纹形核	"Chevron" Markings	人字纹花样
Microcrack Propagation	微裂纹扩展	Screw Dislocation	螺型位错
Zener-Stroh Crack	甄纳-斯特罗裂纹	Quasi Cleavage Fracture	准解理断裂
Pileup Of Dislocations	位错塞积	Transgranular Fracture	穿晶断裂
Micro-Void/Cavity Nucleation	微孔形核	Intergranular Fracture	沿晶断裂
Micro-Void Growth	微孔长大	Shear Fracture	纯剪切断裂
Micro-Void Accumulation/ Coalescence	微孔聚合	Necking Phenomena	缩颈现象
Cup-And-Cone Fracture	杯锥状断裂	Face-Centered Cubic (FCC) Structure	面心立方结构
Fractographic Examinations	断口观察	Body-Centered Cubic (BCC) Structure	体心立方结构
Dimple	韧窝	Hexagonal Close-Packed (HCP) Structure	密排六方结构
Brittle Fracture	脆性断裂	Ductile–Brittle Transition	韧脆转变
Cleavage Fracture	解理断裂	Ductile–Brittle Transition Temperature (Dbtt)	韧脆转变温度
Cleavage Facet	解理面	Charpy Impact Test	摆锤式冲击试验

Chapter 4

Principles of Fracture Mechanics

Learning Objectives

After studying this chapter you should be able to do the following:

1. Explain why the strengths of brittle materials are much lower than predicted by theoretical calculations.

2. Define fracture toughness in terms of (a) a brief statement, and (b) an equation; define all parameters in this equation.

3. Make a distinction between fracture toughness and plane strain fracture toughness.

4. Name and describe the two impact fracture testing techniques.

4.1 Introduction

In this chapter, we will develop a quantitative understanding of cracks. It is very important to calculate the stresses at the tip (or in the vicinity of the tip) of a crack, because these calculations help us answer a very important practical question: At what value of the external load will a crack start to grow?

The significant discrepancy between actual and theoretical fracture strengths of brittle materials is explained by the existence of small flaws that are capable of amplifying an applied tensile stress in their vicinity, leading ultimately to crack formation. Fracture ensues when the theoretical cohesive strength is exceeded at the tip of one of these flaws.

Linear elastic fracture mechanics (LEFM) applies the theory of linear elasticity to the phenomenon of fracture—mainly, the propagation of cracks. If we define the fracture toughness of a material as its resistance to crack propagation, then we can use LEFM to provide us with a quantitative measure of fracture toughness. Various standardization bodies, such as the American Society for Testing and Materials (ASTM), British Standards Institution (BSI), and Japan Institute of Standards (JIS), have standards for fracture toughness tests.

Among the parameters and tests that have been developed, mostly during the last quarter of the twentieth century, to describe the resistance to fracture of a material in a quantitative and reproducible manner, is the plane strain fracture toughness, defined as the critical stress intensity factor under plane strain conditions and mode I loading. This is the stress intensity factor at which a crack of a given size starts to grow in an unstable manner. The fracture toughness is related to the applied stress by an equation of the following form:

$$K_{\mathrm{I}c}=Y\sigma\sqrt{\pi a}$$

where $K_{\mathrm{I}c}$ is the fracture toughness in mode Ⅰ loading, a is the characteristic dimension (semilength) of the crack and Y is a factor that depends on the geometry of the specimen, the location of the crack, and the loading configuration. One can see that the stress which can be safely applied decreases with the square root of the size of the crack. Also, note that K_{Ic} is a parameter of the material in the same manner as are hardness and yield strength. First we derive an expression for the theoretical tensile strength of a crystal. A material is said to cleave when it breaks under normal stress and the fracture path is perpendicular to the applied stress. The process involves the separation of the atoms along the direction of the applied stress. Orowan developed a simple method for obtaining the theoretical tensile strength of a crystal. With his method, no stress concentrations at the tip of the crack are assumed; instead, it is assumed that all atoms separate simultaneously once their separation reaches a critical value.

4.2 Theoretical Cleavage Strength

Figure 4.1 shows how the stress required to separate two planes will vary as a function of the distance between planes. The distance is initially equal to a_0. Naturally, σ for $a = a_0$; σ will also be zero when the separation is infinite. The exact form of the curve of σ versus depends on the nature of the interatomic forces. In Orowan's model, the curve is simply assumed to be a sine function—hence the generality of the model. The area under the curve is the work required to cleave the crystal. This work of deformation—and here there is a certain similarity with Griffith's crack propagation theory to be presented in Section 4.4—cannot be lower than the energy of the two new surfaces created by the cleavage. If the surface energy per unit area is γ and the cross-sectional area of the specimen is A, the total energy is $2\gamma A$ (two surfaces formed). The stress dependence on plane separation is then given by the following equations, admitting a sine function and assuming a periodicity of $2d$:

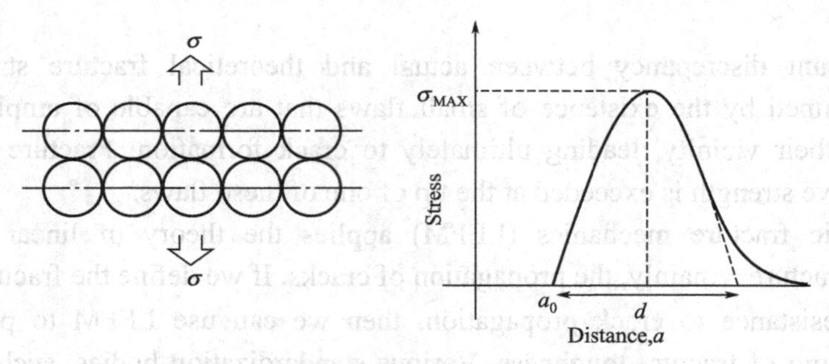

Figure 4.1 Stress required to separate two atomic layers.

$$\sigma = K \sin \frac{2\pi}{2d}(a - a_0) \tag{4.1}$$

K is a constant that can be determined by the following artifice: When a is close to a_0, the material responds linearly to the applied loads (Hookean behavior). Assuming that the elastic deformation is restricted to the two planes shown in Figure 4.2 and that the material is isotropic, the fractional change in the distance between the planes, $\mathrm{d}a/a_0$, is defined as t he incremental strain $\mathrm{d}\varepsilon$.

$$\frac{\mathrm{d}a}{a_0} = \mathrm{d}\varepsilon$$

$$\frac{\mathrm{d}\sigma}{\mathrm{d}\varepsilon} = \frac{\mathrm{d}\sigma}{\mathrm{d}a / a_0} = E \tag{4.2}$$

where E is Young's modulus, which is defined as dσ/dε in the elastic region. Thus,

$$a_0 \frac{d\sigma}{da} = E$$

Taking the derivative of Equation 4.1 and substituting into Equation 4.2 for $a = a_0$,

$$a_0 \frac{\mathrm{d}\sigma}{\mathrm{d}a} = K\frac{\pi}{d} a_0 \cos\frac{\pi}{d}(a - a_0) = E\,,$$

$$K = \frac{E}{\pi} \times \frac{d}{a_0} \tag{4.3}$$

However, d is not known; to determine d, the area under the curve has to be equated to the energy of the two surfaces created:

$$\int_{a_0}^{a_0+d} \sigma \mathrm{d}a = 2\gamma \tag{4.4}$$

Substituting Equation 4.1 into 4.4, we get

$$\int_{a_0}^{a_0+d} K \sin\frac{2\pi}{2d}(a - a_0)\mathrm{d}a = 2\gamma \tag{4.5}$$

From a standard mathematics text, the preceding integral can be evaluated:

$$\int \sin ax\mathrm{d}x = -\frac{1}{a}\cos ax \tag{4.6}$$

A substitution of variables is required to solve Equation 4.5; applying the standard Equation 4.6, we have $a - a_0 = y$; therefore, da = dy, and

$$K\int_0^d \sin\frac{\pi}{d} y\mathrm{d}y = 2\gamma$$

$$K\frac{d}{\pi} = \gamma$$

and

$$d = \frac{\pi\gamma}{K} \tag{4.7}$$

The maximum value of σ is equal to the theoretical cleavage stress. From Equation 4.1, and making the sine equal to 1, we have, from Equation 4.3

$$\sigma_{\max} = K = \frac{E}{\pi} \times \frac{d}{a_0} \tag{4.8}$$

Table 4.1 Theorectical Cleavage Stresses According to Orowan's Theory

Element	Direction	Young's Modulus/GPa	Surface Energy/(J/m^2)	$\sigma_{\max}$/GPa	$\sigma_{\max}/E$
σ-Lron	⟨100⟩	132	2	30	0.23
	⟨111⟩	260	2	46	0.18
Silver	⟨111⟩	121	1.13	24	0.20
Gold	⟨111⟩	110	1.35	27	0.25
Copper	⟨111⟩	192	1.65	39	0.20
	⟨100⟩	67	1.65	25	0.38
Tungsten	⟨100⟩	390	3.00	86	0.22
Diamond	⟨111⟩	1210	5.4	205	0.17

* Adapted with permission from A Kelly. Strong Solids. 2nd ed. (Oxford, U.K.: Clarendon Press,1973), p.73.

Substituting Equation 4.7 into Equation 4.8 yields

$$K = \sigma_{\max} = \frac{E\gamma}{a_0 K}$$

and

$$K^2 = (\sigma_{\max})^2 = \frac{E\gamma}{a_0}$$

or
$$\sigma_{max} = \sqrt{\frac{E\gamma}{a_0}} \tag{4.9}$$

According to Orowan's model, the surface energy is given by

$$\gamma = \frac{Kd}{\pi} = \frac{E}{a_0}\left(\frac{d}{\pi}\right)^2 \tag{4.10}$$

$$\gamma = \frac{Ea_0}{10} \text{ and } \sigma_{max} \approx \frac{E}{\pi} \tag{4.11}$$

We can conclude from Equation 4.7 that, in order to have a high theoretical cleavage strength, a material must have a high Young's modulus and surface energy and a small distance a_0 between atomic planes. Table 4.1 presents the theoretical cleavage strengths for a number of metals. The greatest source of error is γ: it is not easy to determine γ with great precision in solids, and the values used in the table come from different sources and were not necessarily determined at the same temperature.

4.3 Stress Concentration

The most fundamental requisite for the propagation of a crack is that the stress at the tip of the crack must exceed the theoretical cohesive strength of the material. This is indeed the fundamental criterion, but it is not very useful, because it is almost impossible to measure the stress at the tip of the crack. An equivalent criterion, called the Griffith criterion, is more useful and predicts the force that must be applied to a body containing a crack for the propagation of the crack. The Griffith criterion is based on an energy balance and is described in Section 4.4. Let us first grasp the basic idea of stress concentration in a solid.

4.3.1 Stress Concentrations

The failure of a material is associated with the presence of high local stresses and strains in the vicinity of defects. Thus, it is important to know the magnitude and distribution of these stresses and strains around cracklike defects.

Consider a plate having a through-the-thickness notch and subjected to a uniform tensile stress away from the notch (Figure 4.2). We can imagine the applied external force being transmitted from one end of the plate to the other by means of lines of force (similar to the well-known magnetic lines of force). At the ends of the plate, which is being uniformly stretched, the spacing between the lines is uniform. The lines of force in the central region of the plate are severely distorted by the presence of the notch (i.e., the stress field is perturbed). The lines of force, acting as elastic strings, tend to minimize their lengths and thus group together near the ends of the elliptic hole. This grouping together of lines causes a decrease in the line spacing locally and, consequently, an increase in the local stress (a stress concentration), there being more lines of force in the same area.

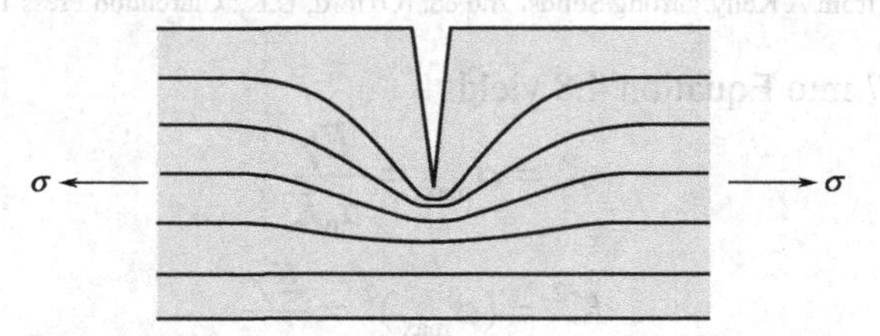

Figure 4.2 "Lines of force" in a bar with a side notch.

4.3.2 Stress Concentration Factor

The theoretical fracture stress of a solid is on the order $E/10$ (see Section 4.2), but the strength of solids (crystalline or otherwise) in practice is orders of magnitude less than this value. The first attempt at giving a rational explanation of this discrepancy was due to Griffith. His analytical model was based on the elastic solution of a cavity elongated in the form of an ellipse.

Figure 4.2 shows "Lines of force" in a bar with a side notch. The direction and density of the lines indicate the direction and magnitude of stress in the bar under a uniform stress σ away from the notch. There is a concentration of the lines of force at the tip of the notch.

Figure 4.3 shows an elliptical cavity in a plate under a uniform stress σ away from the cavity. The maximum stress occurs at the ends of the major axis of the cavity and is given by Inglis's formula

$$\sigma_{\max} = \sigma\left(1+2\frac{a}{b}\right) \tag{4.12}$$

where $2a$ and $2b$ are the major and minor axes of the ellipse, respectively. The value of the stress at the leading edge of the cavity becomes extremely large as the ellipse is flattened. In the case of an extremely flat ellipse or a very narrow crack of length $2a$ and having a radius of curvature $\rho = b^2/a$, Equation 4.12 can be written as

$$\sigma_{\max} = \sigma\left(1+2\sqrt{\frac{a}{b}}\right) \approx 2\sigma\sqrt{\frac{a}{b}} \quad \text{for} \quad \rho \ll a \tag{4.13}$$

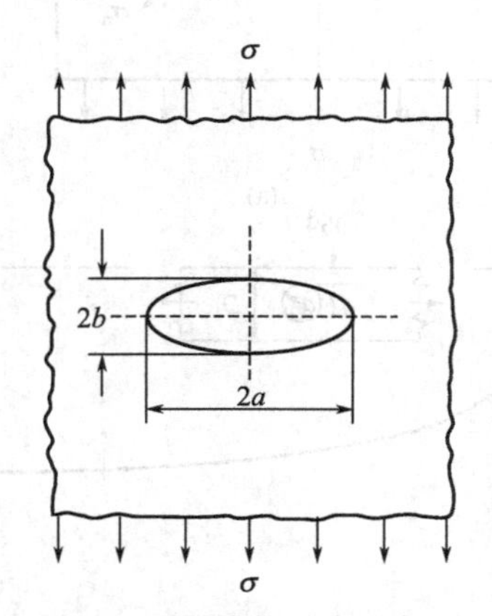

Figure 4.3 Griffith model of a crack.

We note that as ρ becomes very small, $\sigma_{\max}$ becomes very large, and in the limit, as $\rho \to 0$, $\sigma_{\max} \to \infty$. We define the term $2\sqrt{a/\rho}$ as the stress concentration factor K_t (i.e., $K_t = \sigma_{\max}/\sigma$). K_t simply describes the geometric effect of the crack on the local stress (i.e., at the tip of the crack). Note that K_t depends more on the *form* of the cavity than on its size. A number of texts and handbooks give a compilation of stress concentration factors K_t for components containing cracks or notches of various configurations.

As an example of the importance of stress concentration, we point out the use of square windows in the COMET commercial jet aircraft. Fatigue cracks, initiated at the corners of the windows, caused catastrophic failures of several of these aircraft.

In addition to producing a stress concentration, a notch produces a local situation of biaxial or triaxial stress. For example, in the case of a plate containing a circular hole and subject to an axial force, there exist radial as well as tangential stresses. The stresses in a large plate containing a circular hole (with diameter $2a$) and axially loaded [Figure 4.4 (a)] can be expressed as

$$\sigma_{rr} = \frac{\sigma}{2}\left(1-\frac{a^2}{r^2}\right)+\frac{\sigma}{2}\left(1+3\frac{a^4}{r^4}-4\frac{a^2}{r^2}\right)\cos 2\theta$$

$$\sigma_{\theta\theta} = \frac{\sigma}{2}\left(1+\frac{a^2}{r^2}\right)-\frac{\sigma}{2}\left(1+3\frac{a^4}{r^4}\right)\cos 2\theta$$

$$\sigma_{r\theta} = -\frac{\sigma}{2}\left(1-3\frac{a^4}{r^4}+2\frac{a^2}{r^2}\right)\sin 2\theta \tag{4.14}$$

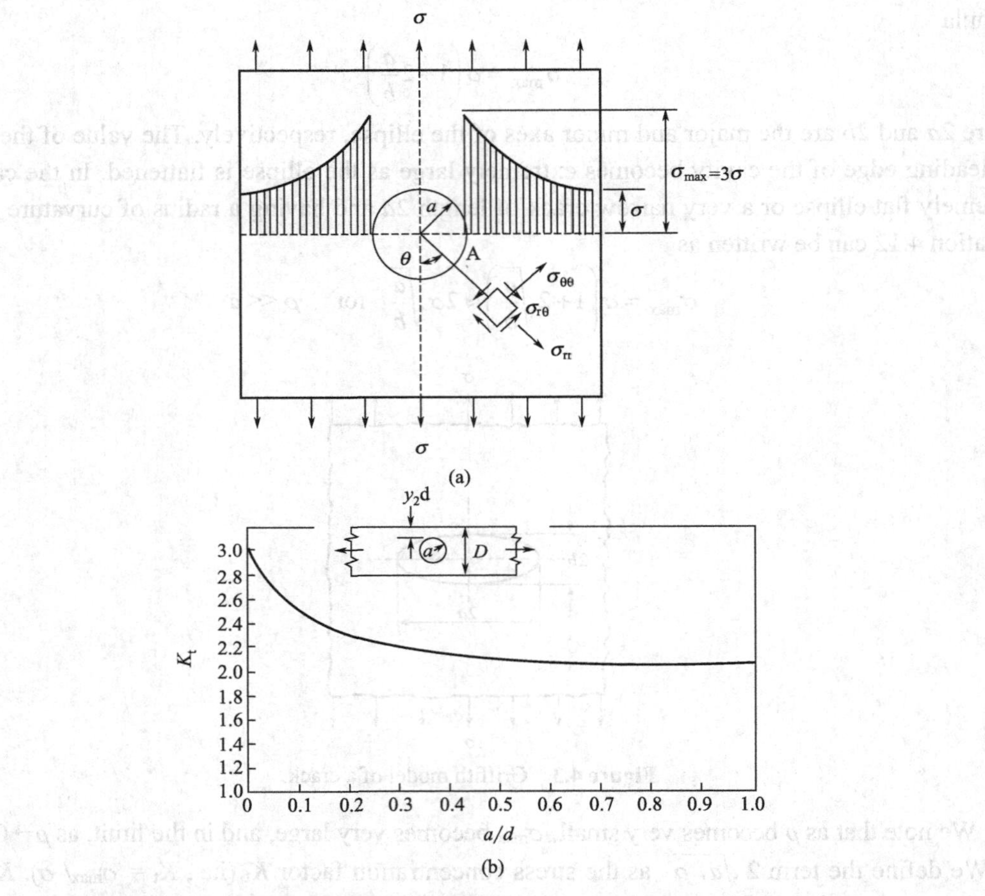

Figure 4.4 (a) Stress distribution in a large plate containing a circular hole.
(b) Stress concentration factor K_t as a function of the radius of a circular hole in a large plate in tension.

The maximum stress occurs at point A in Figure 4.4(a), where $\theta = \pi/2$ and $r = a$. In this case,

$$\sigma_{\theta\theta} = 3\sigma = \sigma_{max}$$

where σ is the uniform stress applied at the ends of the plate. The stress concentration $K_t = \sigma_{max}/\sigma = 3$. Figure 4.4(b) shows the stress concentration for a circular hole in a plate of finite lateral dimensions. When D, the lateral dimension, decreases, or the radius of the hole increases, the stress concentration K_t drops from 3 to 2.2.

Goodier calculated the stresses around spherical voids in perfectly elastic materials. Although his solution was obtained when the applied stress was tensile, it can be extended to compressive stress by changing the signs. The stresses given by Timoshenko and Goodier can be determined from the methods of elasticity theory. At the equatorial plane ($\theta = \pi/2$), the tangential stress $\sigma_{\theta\theta}$ is equal to

$$\sigma_{\theta\theta} = \left[1 + \frac{4-5\nu}{2(7-5\nu)} \times \frac{a^3}{r^3} + \frac{9}{2(7-5\nu)} \times \frac{a^5}{r^5}\right]\sigma \tag{4.15}$$

where a is the radius of the hole, r is the radial coordinate, and ν is the Poisson's ratio. For $r = a$, $\nu =$ 0.3, and we have

$$(\sigma_{\theta\theta})_{\max} \approx 2\,\sigma$$

Thus, as expected, the stress concentration for a spherical void is approximately. The stress $\sigma_{\theta\theta}$ decays quite rapidly with r, as can be seen from Equation 4.15; the decay is given by r^{-3}. For $r = 2a$, we have $\sigma_{\theta\theta} = 1.054$. This decay is faster than for the circular hole, where it goes with r^{-2} (Equation 4.14). For $\theta = 0$ (north and south poles), Timoshenko and Goodier have the equation

$$(\sigma_{rr})_\theta = (\sigma_{\theta\theta})_{\theta=0} = -\frac{3+15\nu}{2(7-5\nu)}\sigma$$

Hence, a compressive stress generates a tensile stress at $\theta = 0$. This result is very important and shows that compressive stress can generate cracks at spherical flaws such as voids. Taking $\nu =$ 0.2~0.3 (typical of ceramics), one arrives at the following values:

$$\frac{1}{2} \leqslant (\sigma_{\theta\theta})_{\theta=0} \leqslant \frac{7.5}{11}$$

Thus, the tensile stress is 50%~80% of the applied compressive stress. If failure is determined by cracking at spherical voids, cracking should start at a compressive stress level equal to $-4t$ (depending on ν; in this case, for $\nu = 0.2$), where σ_t is the tensile strength of the material.

This value represents, to a first approximation, the marked differences between the tensile and compressive strengths of cast irons, intermetallic compounds, and ceramics. The result is fairly close to the stress generated around a circular hole, given in Equation 4.14. In that case, for $r = a$, we find that

$$\sigma_{\theta\theta} = -\,\sigma$$

In tensile loading, the stress $\sigma_{\theta\theta} = 3\,\sigma$, which would predict a three fold difference in tensile and compressive strengths. More general (elliptical) flaws can be assumed, and their response under compressive loading provides a better understanding of the compressive strength of brittle materials. The generation and growth of cracks from these flaws also needs to be analyzed, for more realistic predictions.

Stress concentration caused by an elliptical hole is shown in Figure 4.5. In this figure, σ_L is the longitudinal stress applied along x_2. It is also referred to as the far-field stress. Locally at the crack tip we have a biaxial or triaxial stress situation. In particular, for an elliptical hole, with $a = 3b$, Figure 4.5 shows that σ_{22} falls from its maximum value at the

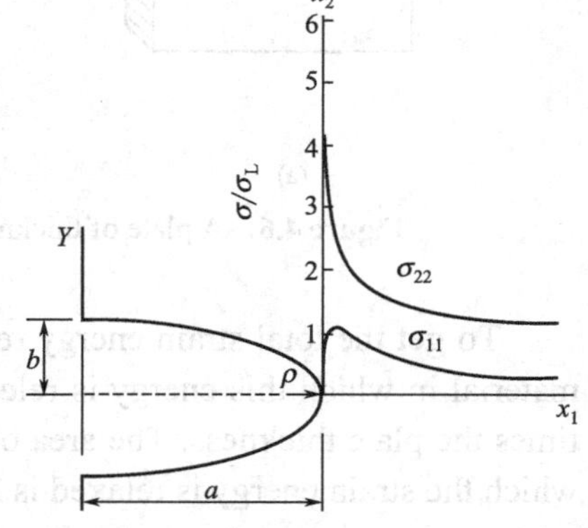

Figure 4.5 Stress concentration at an elliptical hole for $a = 3b$.

crack tip and attains σ_L asymptotically. The stress component, σ_{11}, however is zero at the crack tip, increases to a peak value and then falls to zero with the same tendency as σ_{22}.

The general result is that a major perturbation in the applied stress state occurs over a distance approximately equal to a from the boundaries of the cavity, with the major stress gradients being confined to a region of dimensions roughly equal to ρ surrounding the maximum concentration position.

Although the exact formulas vary according to the form of the crack, in all cases K_t increases with an increase in the crack length a and a decrease in the root radius at the crack tip, ρ.

Despite the fact that the analysis of Inglis represented a great advance, the fundamental nature of the fracture mechanism remained obscure. If the Inglis analysis was applicable to a body containing a crack, how does one explain that, in practice, larger cracks propagate more easily than smaller cracks? What is the physical significance of the root radius at the tip of the crack?

4.4 Griffith Criterion of Fracture

Griffith proposed a criterion based on a thermodynamic energy balance. He pointed out that two things happen when a crack propagates: Elastic strain energy is released in a volume of material, and two new crack surfaces are created, which represent a surface-energy term. Thus, according to Griffith, an existing crack will propagate if the elastic strain energy released by doing so is greater than the surface energy created by the two new crack surfaces. Figure 4.6(a) shows an infinite plate of thickness t that contains a crack of length $2a$ under plane stress. As the stress is applied, the crack opens up. The shaded region denotes the approximate volume of material in which the stored elastic strain energy is released [Figure 4.6(b)]. When the crack extends a distance da on the extremities, the volume over which elastic energy is released increases, as shown in Figure 4.6(c). The elastic energy per unit volume in a solid under stress is given by $\sigma^2/2E$.

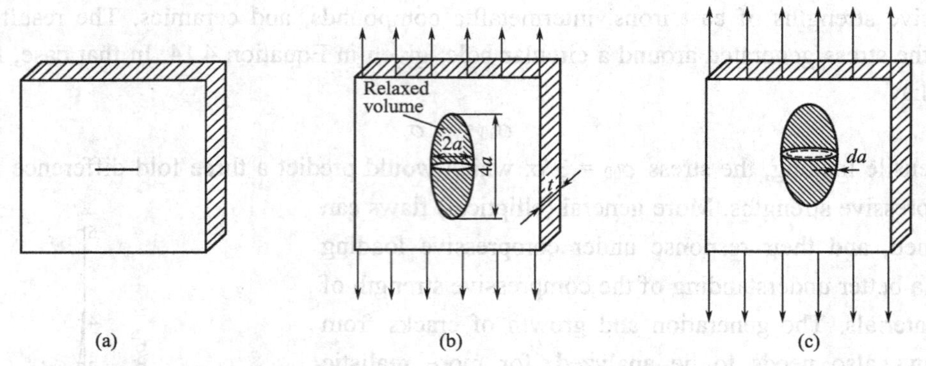

Figure 4.6 A plate of thickness t containing a crack of length $2a$. (a) Unloaded condition. (b) and (c) Loaded condition.

To get the total strain energy released, we need to multiply this quantity by the volume of the material in which this energy is released. In the present case, this volume is the area of the ellipse times the plate thickness. The area of the shaded ellipse is $\pi(2a)a = 2\pi a^2$; therefore, the volume in which the strain energy is relaxed is $2\pi a^2 t$. The total strain energy released is thus

$$\left(\frac{\sigma^2}{2E}\right)\times(2\pi a^2 t)=\frac{\pi\sigma^2 a^2 t}{E}$$

or, in terms of the per-unit thickness of the plate under plane stress, the energy released is

$$U_e = \pi\sigma^2 a^2 / E$$

The decrease in strain energy, U_e, when a crack propagates is balanced by an increase in the surface energy, U_s, produced by the creation of the two new crack surfaces. The increase in surface energy equals:

$$U_s = (2at)(2\gamma_s)$$

here γ_s is the specific surface energy, i.e., the energy per unit area. In terms of the per-unit thickness of the plate, the increase in surface energy is $4a\gamma_s$. Now, when an elliptical crack is introduced into the plate, we can write, for the change in potential energy of the plate,

$$\Delta U = U_s - U_e$$

$$\Delta U = 4a\gamma_s - \frac{\pi\sigma^2 a^2}{E}$$

where ΔU is the change in the potential energy per unit thickness of the plate in the presence of the crack, σ is the applied stress, a is half the crack length, E is the modulus of elasticity of the plate, and γ_s is the specific surface energy (i.e., the surface energy per unit area) of the plate.

As the crack grows, strain energy is released, but additional surfaces are created. The crack becomes stable when these energy components balance each other. If they are not in balance, we have an unstable crack (i.e., the crack will grow). We can obtain the equilibrium condition by equating to zero the first derivative of the potential energy ΔU with respect to the crack length. Thus,

$$\frac{\partial U}{\partial a} = 4\gamma_s - \frac{2\pi\sigma^2 a}{E} = 0 \qquad (4.16a)$$

or

$$2\gamma_s = \frac{\pi\sigma^2 a}{E} \qquad (4.16b)$$

The reader can check the nature of this equilibrium further by taking the second derivative of U with respect to a. A negative second derivative would imply that Equations 4.16a represent an unstable equilibrium condition and that the crack will advance.

Rearranging Equation 4.16b, we may write, for the critical stress required for the crack to propagate in the plane-stress situation,

$$\sigma_c = \sqrt{\frac{2E\gamma_s}{\pi a}} \quad \text{(plane stress)} \qquad (4.17a)$$

We can rearrange Equation 4.17a to get the following expression:

$$\sigma\sqrt{\pi a} = \sqrt{2E\gamma_s}$$

The reader should note that the left-hand side of this expression involves critical stress for crack propagation and square root of crack length. This product is called fracture toughness. Note that the right hand side of the expression consists only of material parameters: E and γ_s, i.e., the above expression represents a material property, viz., fracture toughness.

For the plane-strain situation, we will have the factor $(1 - \nu^2)$ in the denominator because of the confinement in the direction of thickness. The expression for the critical stress for crack propagation then becomes

$$\sigma_c = \sqrt{\frac{2E\gamma_s}{\pi a(1-\nu^2)}} \quad \text{(plane strain)} \qquad (4.17b)$$

The distinction between plane stress and plane strain is shown in Figure 4.7. Normal and shear stresses at free surfaces are zero; hence, for a thin plate, $\sigma_{33} = \sigma_{23} = \sigma_{13} = 0$. This is a plane-stress state [Figure 4.7(a)]. In very thick plates ($t_2 > t_1$), the flow of material in the x_3 direction is restricted. Therefore, $\varepsilon_{33} = 0$, and consequently, $\varepsilon_{23} = \varepsilon_{13} = 0$. This is a plane-strain condition [Figure 4.7(b)]. Note that the factor $(1 - \nu^2)$ is less than unity and is in the denominator. Therefore, the critical stress corresponding to fracture in a plane-strain situation will be higher than that in the plane-stress state.

This is as expected, because of the confinement in the direction of thickness in the case of plane strain. For many metals, $\nu \approx 0.3$, and $(1 - \nu^2) \approx 0.91$. Thus, the difference is not very large for most metals.

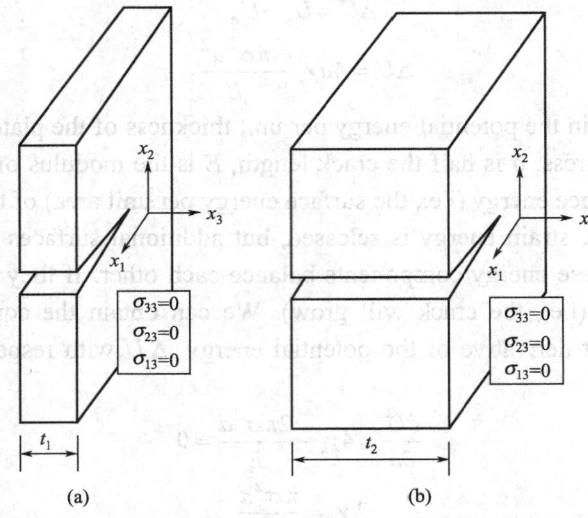

Figure 4.7 Crack in (a) thin (t_1) and (b) thick (t_2) plates. Note the plane-stress state in (a) and the plane-strain state in (b).

Let us consider Equation 4.17a or Equation 4.17b again. Note that the fracture stress, or critical stress required for crack propagation, σ_c, is inversely proportional to $\sqrt{a}$. More importantly, the quantity $\sigma_c\sqrt{a}$ depends only on material constants. It is instructive, then, to examine the Inglis result, Equation 4.13, and the Griffith result, Equation 4.17a or Equation 4.17b in the form

$$\sigma_c\sqrt{a} = \frac{1}{2}(\sigma_{max})_c\sqrt{\rho} = \text{constant}$$

Here, σ_c is the critical far-field or uniform stress (i.e., the stress at fracture), a is the crack length corresponding to σ_c, $(\sigma_{max})_c$ is the stress at the crack tip at fracture, and ρ is the root radius at the tip of the crack.

Both analyses, Inglis's and Griffith's lead to the same result, viz., that a crack will propagate when an appropriate quantity with dimensions of stress times the square root of length reaches a critical value, a material constant. It is easy to see that the parameters in the Inglis analysis, $(\sigma_{max})_c$ and ρ, are local parameters and very difficult to measure, while the Griffith analysis allows us to use the far-field applied stress and crack length, which are easy to measure. It is this quantity, $\sigma_c\sqrt{a}$, that is called the fracture toughness and is denoted by K_{Ic}.

If the material in which a crack is propagating can deform plastically, the form of the crack tip changes because of plastic strain. A sharp crack tip will be blunted. Another important factor is time:

Because plastic deformation requires time, the amount of plastic deformation that can occur at the crack tip will depend on how fast the crack is moving. In a great majority of materials, localized plastic deformation at and around the crack tip is produced because of the stress concentrations there. In such a case, a certain amount of plastic work is done during crack propagation, in addition to the elastic work done in the creation of two fracture surfaces. The mechanics of fracture will, then, depend on the magnitude of γ_p, the plastic work done, which in its turn depends on the crack speed, temperature, and the nature of the material. For an inherently brittle material, at low temperatures and at high crack velocities γ_p is relatively small ($\gamma_p < 0.1\gamma_s$). In such a case, the crack propagation would be continuous and elastic. These cases are usefully treated by means of linear elastic fracture mechanics, which is dealt with in Section 4.6. In any event, in the case of plastic deformation, the work done in the propagation of a crack per unit area of the fracture surface is increased from γ_s to $(\gamma_s + \gamma_p)$. Consequently, the Griffith criterion (Equation 4.17a or Equation 4.17b) is modified to

$$\sigma_c = \sqrt{\frac{2E(\gamma_s + \gamma_p)}{\pi a}} \quad \text{(plane stress)} \qquad (4.18a)$$

and

$$\sigma_c = \sqrt{\frac{2E(\gamma_s + \gamma_p)}{\pi a(1-\nu^2)}} \quad \text{(plane strain)} \qquad (4.18b)$$

Rearranging Equation 4.18a, we get

$$\sigma_c = \sqrt{\frac{2E\gamma_p}{\pi a}\left(1 + \frac{\gamma_s}{\gamma_p}\right)}$$

For $\gamma_p/\gamma_s \gg 1$,

$$\sigma_c = \sqrt{\frac{2E\gamma_p}{\pi a}}$$

Thus, the plastic deformation around the crack tip makes it blunt and serves to relax the stress concentration by increasing the radius of curvature of the crack at its tip. Localized plastic deformation at the crack tip therefore improves the fracture toughness of the material.

This is the conventional treatment of the plastic work contribution to the fracture process, wherein γ_p is considered to be a constant. However, the reader should be warned that this is not strictly true. As a matter of fact, the value of γ_p increases with the stress intensity factor K $(=Y\sigma\sqrt{a})$.

As pointed out earlier, equations of the type 4.17a or 4.18 are difficult to use in practice. It is not a trivial matter to measure quantities such as surface energy and the energy of plastic deformation. In a manner similar to that of Griffith, Irwin made a fundamental contribution to the mechanics of fracture when he proposed that fracture occurs at a stress that corresponds to a critical value of the crack extension force

$$G = \frac{1}{2} \times \frac{\partial U_e}{\partial a} = \text{rate of change of energy with crack length.}$$

G is sometimes called the *strain energy release rate*.

Now, $U_e = \pi a^2 \sigma^2/E$, the energy released by the advancing crack per unit of plate thickness. This is for plane stress. For plane strain, a factor of $(1 - \nu^2)$ is introduced in the denominator. Thus,

$$G = \frac{\pi a \sigma^2}{E}$$

At fracture, $G = G_c$, and

$$\sigma_c = \sqrt{\frac{EG_c}{\pi a}} \quad \text{(plane stress)} \tag{4.19a}$$

or $$\sigma_c = \sqrt{\frac{EG_c}{\pi a(1-\nu^2)}} \quad \text{(plane strain)} \tag{4.19b}$$

From Equation 4.18 and Equation 4.19, we see that

$$G_c = 2(\gamma_s + \gamma_p)$$

We shall come back to this idea of crack extension force later in the chapter.

EXAMPLE PROBLEM 4.1

Maximum Flaw Length Computation

A relatively large plate of a glass is subjected to a tensile stress of 40MPa. If the specific surface energy and modulus of elasticity for this glass are 0.3J/m^2 and 69GPa, respectively, determine the maximum length of a surface flaw that is possible without fracture.

Solution

To solve this problem it is necessary to employ Equation 3.3. Rearrangement of this expression such that a is the dependent variable, and realizing that $\sigma = 40\text{MPa}$, $\gamma_s = 0.3\text{J/m}^2$ and $E = 69\text{GPa}$ leads to

$$a = \frac{2E\gamma_s}{\pi a^2} = \frac{2\times(69\times10^9\,\text{N/m}^2)\times(0.3\text{N/m})}{\pi(40\times10^6\,\text{N/m}^2)^2} = 8.2\times10^{-6}\,\text{m} = 0.0082\text{mm} = 8.2\mu\text{m}$$

4.5 Fracture Toughness

A nonductile material has a very low capacity to deform plastically; that is, it is not capable of relaxing peak stresses at cracklike defects.

In such a material, a crack will propagate very rapidly with little plastic deformation around the crack tip, resulting in what is called a brittle fracture. Typically, such a fracture is also characterized by a crack propagation that is sudden, rapid, and unstable. In practical terms, this definition of brittleness, which refers to the onset of instability under an applied stress smaller than the stress corresponding to plastic yielding of the material, is very useful. Numerous brittle fractures have occurred in service, and there are abundant examples of them in a great variety of structural and mechanical engineering fields involving ships, bridges, pressure vessels, oil ducts, turbines, and so on. In view of the great importance of brittle fracture in real life, a discipline called linear elastic fracture mechanics (LEFM) has emerged, enabling us to obtain a quantitative measure of the resistance of a brittle material to unstable or catastrophic crack propagation. Extension of these efforts into nonlinear elastic and plastic regimes has led to the development of elastic-plastic fracture mechanics (EPFM), also called post-yield fracture mechanics.

Fracture mechanics gives us a quantitative handle on the process of fracture in materials. Its approach is based on the concept that the relevant material property, fracture toughness, is the force necessary to extend a crack through a structural member. Under certain circumstances, this crack extension force (or an equivalent parameter) becomes independent of the dimensions of the specimen. The parameter can then be used as a quantitative measure of the fracture toughness of the material. Fracture mechanics adopts an entirely new approach to designing against fracture. Admittedly defects will always be present in a structural component. But consider a structure or a component with a cracklike defect. We can simulate this with single edge notch of length a in a plate (Figure 4.8). Alternatively, we can say that we are increasing the applied stress intensity factor K at the crack tip. The material at the tip, however, presents resistance to crack growth. We denote this inherent material resistance by K_R (sometimes the symbol R alone is used in place of K_R). The discipline of fracture mechanics can then be represented by a triangle as shown in Figure 4.8; that is, we have an interplay among the following three quantities:

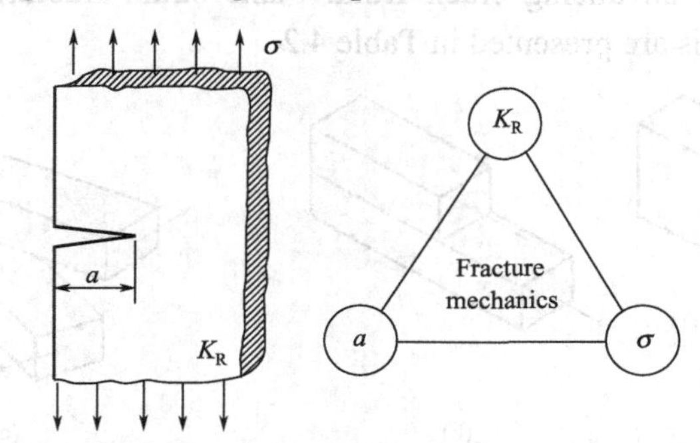

Figure 4.8 Inherent material resistance to crack growth and its relationship to the applied stress σ and crack size a.

(1) The far-field stress, σ.

(2) The characteristic crack length, a.

(3) The inherent material resistance to cracking, K_R.

Various parameters are used to represent K_R. Here we wish to clarify one common point of confusion. The symbol K is used to designate the stress intensity factor at the crack tip corresponding to a given applied stress and crack length. The symbol K_R (or one of its equivalents) represents fracture toughness. In this regard, the following analogy is helpful. The stress intensity factor, K, is to stress as fracture toughness, K_R, is to strength. Stress and stress intensity factor vary with the external loading conditions; strength and toughness are material parameters, independent of loading and specimen size considerations.

4.5.1 Hypotheses of LEFM

The basic hypotheses of LEFM are as follows:

(1) Cracks are inherently present in a material, because there is a limit to the sensibility or resolutions of any crack-detecting equipment.

(2) A crack is a free, internal, plane surface in a linear elastic stress field. With this hypothesis, linear elasticity furnishes us stresses near the crack tip as

$$\sigma_{r\theta}=\frac{K}{\sqrt{2\pi r}}f(\theta) \tag{4.20}$$

where r and θ are polar coordinates and K is a constant called the stress intensity factor (SIF).

(3) The growth of the crack leading to the failure of the structural member is then predicted in terms of the tensile stress acting at the crack tip. In other words, the stress situation at the crack tip is characterized by the value of K. It can be shown by elasticity theory that $K = Y\sigma\sqrt{\pi a}$, where σ is the applied stress, a is half the crack length, and Y is a constant that depends on the crack opening mode and the geometry of the specimen.

4.5.2 Crack-Tip Separation Modes

The three modes of fracture are shown in Figure 4.9. Mode Ⅰ [Figure 4.9(a)], called the opening mode, has tensile stress normal to the crack faces. Mode Ⅱ [Figure 4.9(b)] is called the sliding mode or the forward shear mode. In this mode, the shear stress is normal to the advancing crack front. Mode Ⅲ [Figure 4.9(c)] is called the tearing mode or transverse shear mode, with the shear stress parallel to the advancing crack front. Plane strain fracture toughness values for a number of different materials are presented in Table 4.2.

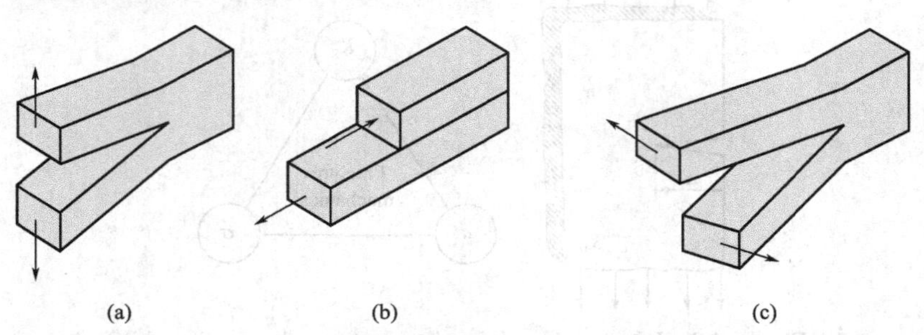

Figure 4.9 The three modes of fracture. (a) Mode Ⅰ: opening mode; (b) mode Ⅱ: sliding mode; and (c) mode Ⅲ: tearing mode.

Table 4.2 Room-Temperature Yield Strength and Plane Strain Fracture Toughness Data for Selected Engineering Materials

Material	Yield Strength MPa	Yield Strength ksi	K_{Ic} MPa $\sqrt{m}$	K_{Ic} ksi $\sqrt{in}$
Metals				
Aluminum Alloy[①](7075-T651)	495	72	24	22
Aluminum Alloy[①](2024-T3)	345	50	44	40
Titanium Alloy[①](Ti-6Al-4V)	910	132	55	50
Alloy Steel[①](4340 tempered at 260°C)	1640	238	50.0	45.8
Alloy Steel[①](4340 tempered at 425°C)	1420	206	87.4	80.0
Ceramics				
Concrete	—	—	0.2~1.4	0.18~1.27
Soda-Lime Glass	—	—	0.7~0.8	0.64~0.73
Aluminum Oxide	—	—	2.7~5.0	2.5~4.6
Polymers				
Polystyrene(PS)	—	—	0.7~1.1	0.64~1.0
Poly(methyl methacrylate)(PMMA)	53.8~73.1	7.8~10.6	0.7~1.6	0.64~1.5
Polycarbonate(PC)	62.1	9.0	2.2	2.0

① Source: Reprinted with permission, *Advanced Materials and Processes*, ASM International, 1990.

4.5.3 Stress Field in an Isotropic Material in the Vicinity of a Crack Tip

The stress components for the three fracture modes in an isotropic material are given next. In the case of anisotropic materials, these relations must be modified to permit the asymmetry of stress at the crack tip. K_{I}, K_{II}, and K_{III} represent stress intensity factors in mode Ⅰ, mode Ⅱ, and mode Ⅲ, respectively. We have (the derivation of these expressions is due to Westergaard):

Mode Ⅰ:

$$\begin{bmatrix}\sigma_{11}\\ \sigma_{22}\\ \sigma_{12}\end{bmatrix}=\frac{K_{\mathrm{I}}}{\sqrt{2\pi r}}\cos\frac{\theta}{2}\begin{bmatrix}1-\sin\frac{\theta}{2} & \sin\frac{3\theta}{2}\\ 1+\sin\frac{\theta}{2} & \sin\frac{3\theta}{2}\\ \sin\frac{\theta}{2} & \cos\frac{3\theta}{2}\end{bmatrix}$$

$$\sigma_{13}=\sigma_{23}=0$$

$$\sigma_{33}=0,\ \text{(plane stress)}$$

$$\sigma_{33}=\nu(\sigma_{11}+\sigma_{22}),\ \text{(plane strain)} \tag{4.21}$$

Mode Ⅱ:

$$\begin{bmatrix}\sigma_{11}\\ \sigma_{22}\\ \sigma_{12}\end{bmatrix}=\frac{K_{\mathrm{II}}}{\sqrt{2\pi r}}\begin{bmatrix}-\sin\frac{\theta}{2}\left(2\cos\frac{\theta}{2}-\cos\frac{3\theta}{2}\right)\\ \sin\frac{\theta}{2}\cos\frac{\theta}{2}\cos\frac{3\theta}{2}\\ \cos\frac{\theta}{2}\left(1-\sin\frac{\theta}{2}\sin\frac{3\theta}{2}\right)\end{bmatrix}$$

$$\sigma_{13}=\sigma_{23}=0,$$

$$\sigma_{33}=0,\ \text{(plane stress)}$$

$$\sigma_{33}=\nu(\sigma_{11}+\sigma_{22})\ \text{(plane strain)} \tag{4.22}$$

Mode Ⅲ:

$$\begin{bmatrix}\sigma_{13}\\ \sigma_{23}\end{bmatrix}=\frac{K_{\mathrm{III}}}{\sqrt{2\pi r}}\begin{bmatrix}-\sin\frac{\theta}{2}\\ \cos\frac{\theta}{2}\end{bmatrix}$$

$$\sigma_{11}=\sigma_{22}=\sigma_{33}=\sigma_{12}=0 \tag{4.23}$$

4.5.4 Details of the Crack-Tip Stress Field in Mode Ⅰ

Consider an infinite, homogeneous, elastic plate containing a crack of length $2a$ (Figure 4.10). The plate is subjected to a tensile stress σ far away from and normal to the crack. The stresses at a point (r, θ) near the tip of the crack are given by Equation 4.21. Ignoring the subscript of K, we may write the stress components in expanded form as:

$$\sigma_{11}=\frac{K}{\sqrt{2\pi r}}\cos\frac{\theta}{2}\left(1-\sin\frac{\theta}{2}\sin\frac{3\theta}{2}\right)$$

$$\sigma_{22}=\frac{K}{\sqrt{2\pi r}}\cos\frac{\theta}{2}\left(1+\sin\frac{\theta}{2}\sin\frac{3\theta}{2}\right)$$

$$\sigma_{12}=\frac{K}{\sqrt{2\pi r}}\cos\frac{\theta}{2}\sin\frac{\theta}{2}\cos\frac{3\theta}{2}$$

$$\sigma_{13}=\sigma_{13}=0$$

$$\sigma_{33}=0\ \text{(plane stress)} \tag{4.24}$$

$$\sigma_{33}=\nu(\sigma_{11}+\sigma_{22})\ \text{(plane strain)}$$

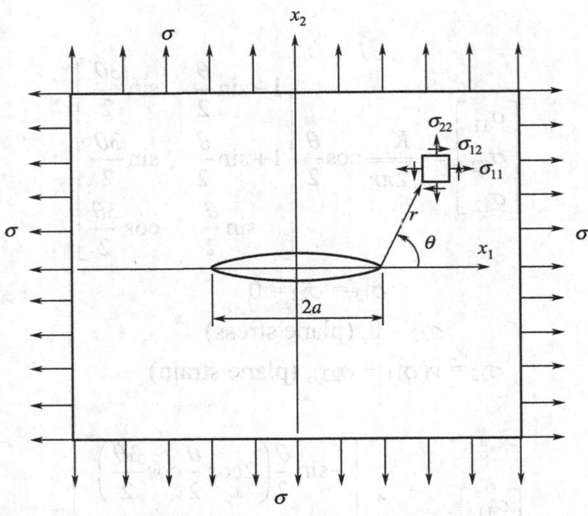

Figure 4.10 Infinite, homogeneous, elastic plate containing a through-the-thickness central crack of length $2a$, subjected to a tensile stress σ.

where

$$K = \sigma\sqrt{\pi a} \tag{4.25}$$

is the stress intensity factor for the plate and has the units $(N/m^2)\sqrt{m}$, or Pa$\sqrt{m}$, or N • $m^{-3/2}$. Note that Equation 4.25 is applicable in the region $r \ll a$ (i.e., in the vicinity of the crack tip). For larger r, higher order terms must be included. For a thin plate, one has plane-stress conditions, and $\sigma_{33} = \sigma_{13} = \sigma_{23} = 0$. For a thick plate (infinite in the direction of thickness), there exist plane-strain conditions [i.e., $\sigma_{33} = \nu(\sigma_{11} + \sigma_{22})$ and $\sigma_{13} = \sigma_{23} = 0$].

Consider again Equation 4.24. The right-hand side has three quantities: K, r, and $f(\theta)$. The terms r and $f(\theta)$ describe the stress distribution around the crack tip. These two characteristics [i.e., dependence on $\sqrt{r}$ and $f(\theta)$] are identical for all cracks in two- or three dimensional elastic solids. The stress intensity factor K includes the influence of the applied stress σ and the appropriate crack dimensions, in this case half the crack length a. Thus, K will characterize the external conditions (i.e., the nominal applied stress σ and half the crack length a) that correspond to fracture when stresses and strains at the crack tip reach a critical value. This critical value of K is designated as K_c. It turns out, as we shall see later, that K_c depends on the dimensions of the specimen. In the case of a thin sample (plane-stress conditions), K_c depends on the thickness of the sample, whereas in the case of a sufficiently thick sample (plane-strain conditions), K is independent of the thickness of the specimen and is designated as $K_{\mathrm{I}c}$.

The stress intensity factor K measures the amplitude of the stress field around the crack tip and should not be confused with the stress concentration factor K_t discussed in Section 4.3.2. It is also important to distinguish between K and K_c or $K_{\mathrm{I}c}$. The stress intensity factor K is a quantity, determined analytically or not, that varies as a function of configuration (i.e., the geometry of the crack and the manner of application of the external load). Thus, the analytical expression for K varies from one system to another. However, once K attains its critical value, $K_{\mathrm{I}c}$, in plane strain for a given system and material, it is essentially a constant for all the systems made of this material. The difference between K_c and $K_{\mathrm{I}c}$ is that K_c depends on the thickness of the specimen, whereas $K_{\mathrm{I}c}$ is independent of the thickness. The forms of K for various load and crack configurations have been calculated and are available in various handbooks. Some of the more common configurations and the corresponding expressions for K are presented in Figure 4.11.

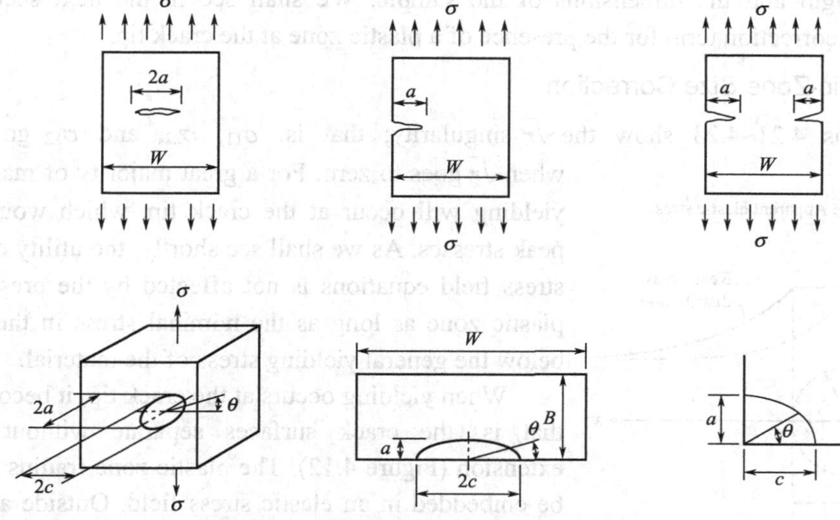

Figure 4.11 Some common load and crack configurations and the corresponding expressions for the stress intensity factor, K.

For samples of finite dimensions, the general practice is to consider the solution for an infinite plate and modify it by an algebraic or trigonometric function that would make the surface tractions vanishes.

Thus, for a central through-the-thickness crack of length $2a$, in a plate of width W, we have

$$K=\sigma\left(W\tan\frac{\pi a}{W}\right)^{1/2} \tag{4.26}$$

For the same crack in an infinite plate, we have $K=\sigma\sqrt{\pi a}$.

If we expand $\tan \pi a/W$ in a series (Equation 4.26), we get

$$K=\sigma W^{1/2}\left(\frac{\pi a}{W}+\frac{\pi^3 a^3}{3W^3}+\cdots\right)^{1/2}$$

$$=\sigma\sqrt{\pi a}\left(1+\frac{\pi^3 a^3}{3W^3}+\cdots\right)^{1/2}$$

Thus, for an infinite solid, $a/W=0$, and we have $K=\sigma\sqrt{\pi a}$, as expected. For an edge crack in a semi-infinite plate, we have $K=1.12\ \sigma\sqrt{\pi a}$. The factor 1.12 here takes care of the fact that stresses normal to the free surface must be zero.

At this point, it is appropriate to make some comments on the limitations of LEFM. It was pointed out earlier that the expressions for stress components (Equations 4.21~4.23) are valid only in the neighborhood of the crack tip. The reader will have noticed that these stress components tend to infinity as we approach the tip (i.e., as r goes to zero). Now, there does not exist a material in real life that can resist an infinite stress. The material in the neighborhood of the crack tip, in fact, would inevitably deform plastically. Thus, these expressions for stress components based on linear elasticity theory are not valid in the plastic zone at the crack tip. The deformation process in a plastic zone, as is well known, will be a sensitive function of the microstructure, among other things. However, in spite of ignorance of the exact nature of the plastic zone, the LEFM treatment is valid for low-enough stresses such that the size of the plastic zone at the crack tip is small with respect to the crack length and the dimensions of the sample. We shall see in the next section how to incorporate a correction term for the presence of a plastic zone at the crack tip.

4.5.5 Plastic-Zone Size Correction

Equations 4.21~4.23 show the $\sqrt{r}$ singularity; that is, σ_{11}, σ_{22}, and σ_{12} go to infinity when $\sqrt{r}$ goes to zero. For a great majority of materials, local yielding will occur at the crack tip, which would relax the peak stresses. As we shall see shortly, the utility of the elastic stress field equations is not affected by the presence of this plastic zone as long as the nominal stress in the material is below the general yielding stress of the material.

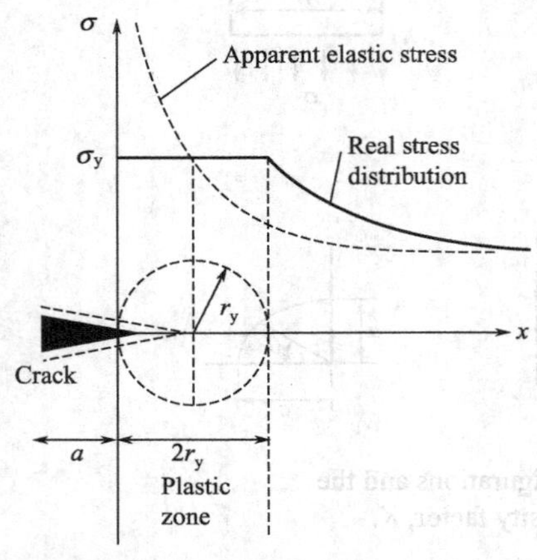

Figure 4.12 Plastic-zone correction. The effective crack length is $(a + r_y)$.

When yielding occurs at the crack tip, it becomes blunted; that is, the crack surfaces separate without any crack extension (Figure 4.12). The plastic zone (radius r_y) will then be embedded in an elastic stress field. Outside and far away from the plastic zone, the elastic stress field "sees" the crack and the perturbation due to the plastic zone, as if there were present a crack in an elastic material with the leading edge of the crack situated inside the plastic zone. A crack of length $2(a + r_y)$ in an ideal elastic material produces stresses almost identical to elastic stresses in a locally yielded member outside the plastic zone. If the stress applied is too large, the plastic zone increases in size in relation to the crack length, and the elastic stress field equations lose precision. When the whole of the reduced section yields, the plastic zone spreads to the edges of the sample, and K does not have any validity as a parameter defining the stress field.

When the plastic zone is small in relation to the crack length, it can be visualized as a cylinder (Figure 4.12) of radius r_y at the crack tip. From Equation 4.24, for $\theta = 0$, $r = r_y$, and $\sigma_{22} = \sigma_y$, the yield stress, we can write

$$\sigma_y = \frac{K}{\sqrt{2\pi r_y}}$$

and, to a first approximation, the plastic-zone radius will be

$$r_y = \frac{1}{2\pi}\left(\frac{K}{\sigma_y}\right)^2 \tag{4.27}$$

In fact, the plastic-zone radius is a little bigger than $(1/2\pi)(K/\sigma_y)^2$, due to redistribution of load in the vicinity of the crack tip. Irwin, taking into account the plastic constraint factor in the case of plane strain, gave the following expressions for the size of the plastic zone:

$$r_y \approx \frac{1}{2\pi}\left(\frac{K}{\sigma_y}\right)^2 \quad \text{(plane stress)}$$

$$r_y \approx \frac{1}{6\pi}\left(\frac{K}{\sigma_y}\right)^2 \quad \text{(plane strain)}$$

Thus, the center of perturbation, the apparent crack tip, is located a distance r_y from the real crack tip. The effective crack length is, then,

$$(2a)_{eff} = 2(a + r_y)$$

Substituting $(a + r_y)$ for a in the elastic stress field equations gives an adequate adjustment for the crack-tip plasticity under conditions of small-scale yielding. With this adjustment, the stress intensity factor K is useful for characterization of the fracture conditions.

There is another model for the plastic zone at the crack tip for the plane-stress case, called the Dugdale–BCS model. In this model, the plasticity spreads out at the two ends of a crack in the form of narrow strips of length R (Figure 4.13). These

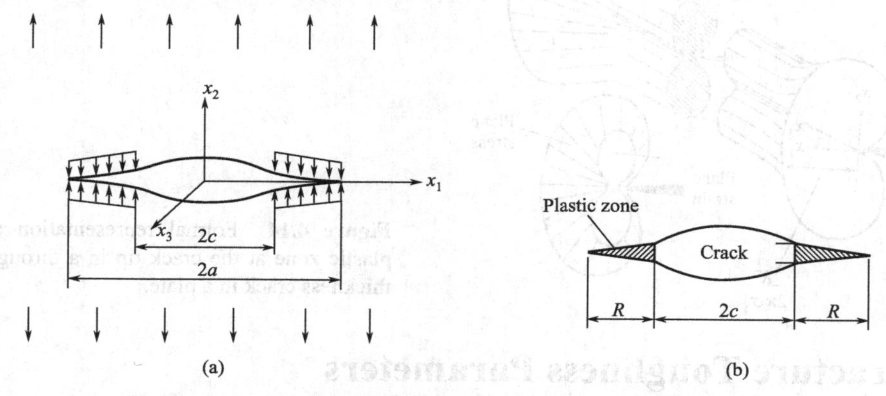

Figure 4.13 Dugdale–Bilby–Cottrell–Swinden model of a crack.

narrow plastic strips in front of the actual crack tips are under the yield stress σ_y that tends to close the crack. Mathematically, the internal crack of length $2c$ is allowed to extend elastically a distance $2a$, and then internal stress is applied to reclose the crack in this region. Combining the internal stress field surrounding the plastic enclaves with the external stress field associated with the applied stress σ acting on the crack, Dugdale showed that

$$\frac{c}{a} = \cos\frac{\pi\sigma}{2\sigma_y}$$

From this relation, one notes that as $\sigma \to \sigma_y$, $c/a \to 0$, $a \to \infty$ (i.e., general yielding occurs). On the other hand, as σ/σ_y decreases, we can write (using the series expansion for cosine),

$$\frac{c}{a} = 1 - \frac{\pi^2\sigma^2}{8\sigma_y^2} + \cdots$$

Noting that $a = c + R$ and using the binomial expansion, we have

$$\frac{c}{a} = \frac{c}{c+R} = \left(1 + \frac{R}{c}\right)^{-1} = 1 - \frac{R}{c} + \cdots$$

Thus, for $\sigma << \sigma_y$,

$$\frac{R}{c} \approx \frac{\pi^2}{8}\left(\frac{\sigma}{\sigma_y}\right)^2$$

or

$$R \approx \frac{\pi}{8}\left(\frac{K}{\sigma_y}\right)^2 \quad \text{(plane stress)} \tag{4.28a}$$

$$R \approx \frac{\pi}{8}\left[\frac{K}{\sigma_y/(1-2\nu)}\right]^2 \quad \text{(plane strain)} \tag{4.28b}$$

Comparing Equation 4.28 with Equation 4.27, we see that there is good agreement between the two ($\pi/8 \approx 1/\pi$). In fact, the size of the plastic zone varies with θ also. A formal representation of the plastic zone at the crack front through the plate thickness is shown in Figure 4.14.

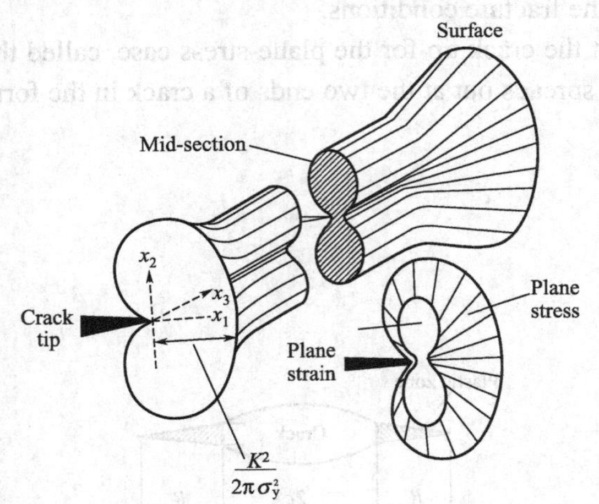

Figure 4.14 Formal representation of the plastic zone at the crack tip in a through-the-thickness crack in a plate.

4.6 Fracture Toughness Parameters

In this section, we describe the variety of fracture toughness parameters that have come into being.

4.6.1 Crack Extension Force G

The concept of the crack extension force G, due to Irwin, can be interpreted as a generalized force. One can say that fracture mechanics is the study of the response of a crack (measured in terms of its velocity) to the application of various magnitudes of the crack extension force. Let us consider an elastic body of uniform thickness B, containing a through-the-thickness crack of length $2a$. Let the body be loaded as shown in Figure 4.15(a). With increasing load P, the displacement e of the loading point increases. The load—displacement diagram is shown in Figure 4.15(b).

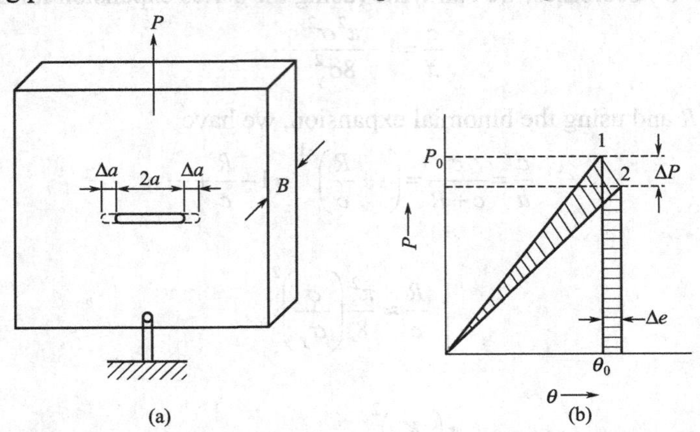

Figure 4.15 (a) Elastic body containing a crack of length $2a$ under load P. (b) Diagram of load P versus displacement e.

At point 1, we have the load as P_0 and displacement as e_0. Now let us consider a "gedanken" experiment in which the crack extends by a small increment, Δa. Due to this small increment in crack extension, the loading point is displaced by Δe, while the load falls by ΔP. Now, before the crack extension, the potential energy stored in body was

$$U_1 = \frac{1}{2} Pe$$

represented by the area of the triangle through point 1 in the figure.

After the crack extension, the potential energy stored in the body is

$$U_2 = \frac{1}{2}(P - \Delta P)(e + \Delta e)$$

represented by the area of the triangle passing through point 2 in the figure. In this process of crack extension, the change in potential energy, $U_2 - U_1$ is given by the difference in the areas of the two crosshatched regions in the figure. Considering the small increment Δa in crack length, we can write an equation for G, the crack extension force per unit length, as

$$GB\Delta a = U_2 - U_1 = \Delta U$$

The change in elastic strain energy with respect to the crack area, in the limit of the area going to zero, equals the crack extension force; that is,

$$G = \lim_{\Delta A \to 0} \frac{\Delta U}{\Delta A}$$

where $\Delta A = B\Delta a$.

It is convenient to evaluate G in terms of the compliance c of the sample, defined as

$$e = cP \tag{4.29}$$

Now,

$$\Delta U = U_2 - U_1 = \frac{1}{2}(P - \Delta P)(e + \Delta e) = \frac{1}{2} Pe \tag{4.30}$$

or

$$\Delta U = \frac{1}{2} P\Delta e - \frac{1}{2} e\Delta P - \frac{1}{2}\Delta P\Delta e \tag{4.31}$$

Differentiating Equation 4.30, we have

$$\Delta e = c\Delta P + P\Delta c \tag{4.32}$$

Substituting Equation 4.32 in Equation 4.31, we obtain

$$\Delta U = \frac{1}{2} Pc\delta P + \frac{1}{2} P^2\Delta c - \frac{1}{2} e\Delta P - \frac{1}{2} e(\Delta P)^2 - \frac{1}{2} P\Delta P\Delta c \tag{4.33}$$

Remembering that $e = cP$ and ignoring the higher order product terms, we can write

$$\Delta U = \frac{1}{2} Pc\Delta P + \frac{1}{2} P^2\Delta c - \frac{1}{2} Pc\Delta c$$

or

$$\Delta U = \frac{1}{2} P^2\Delta c \tag{4.34}$$

Then

$$G = \lim_{\Delta A \to 0} \frac{\Delta U}{\Delta A} = \lim_{\Delta A \to 0} \frac{\frac{1}{2} P^2\Delta c}{\Delta A}$$

or

$$G=\frac{1}{2}\times\frac{P^2}{B}\times\frac{\Delta c}{\Delta a} \tag{4.35}$$

From Equation 4.35, we see that G is independent of the rigidity of the surrounding structure and the test machine. In fact, G depends only on the change in compliance of the cracked member due to crack extension. Thus, to obtain G for a specimen, all we need to do is to determine the compliance of the specimen as a function of crack length and measure the gradient of the resultant curve, dc/da, at the appropriate initial crack length (Figure 4.16).

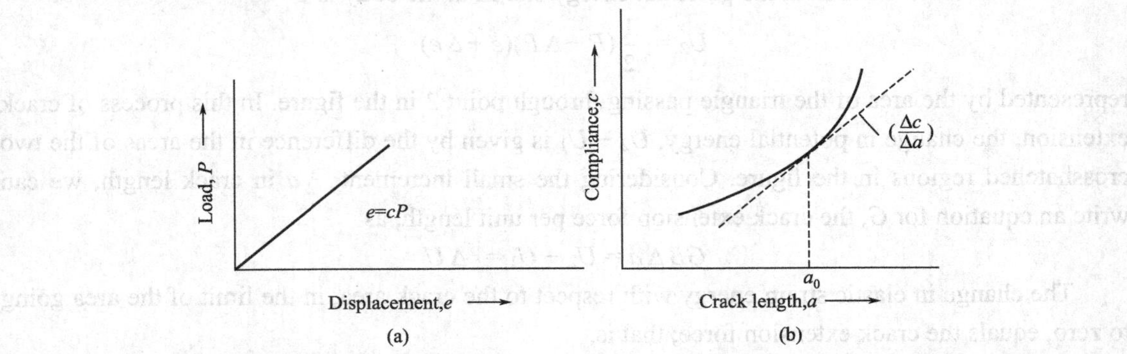

Figrue 4.16 (a) Load P versus displacement e. Compliance c is the inverse of the slope of this curve. (b) Compliance c versus crack length a. a_0 is the initial crack length.

This method is more useful for relatively small test samples, on which exact measurements can be made in the laboratory. One of the important uses of Equation 4.34 is that it provides a value of G (or K) for complex structures that have not been (or cannot be) treated analytically. An experimental determination of G_c, the critical crack extension force, using this equation requires the value of fracture load (measured experimentally) and the value of dc/da. The compliance can be measured by calibrating a series of samples with different crack lengths. We obtain a diagram of c versus a, and dc/da is evaluated as the slope at the appropriate initial crack length.

4.6.2 Crack Tip Opening Displacement (CTOD)

The development of a plastic zone at the tip of the crack results in a displacement of the faces without crack extension. This relative displacement of opposite crack edges is called the crack opening displacement (COD) (Figure 4.17). The crack-tip opening displacement (CTOD), δ, is

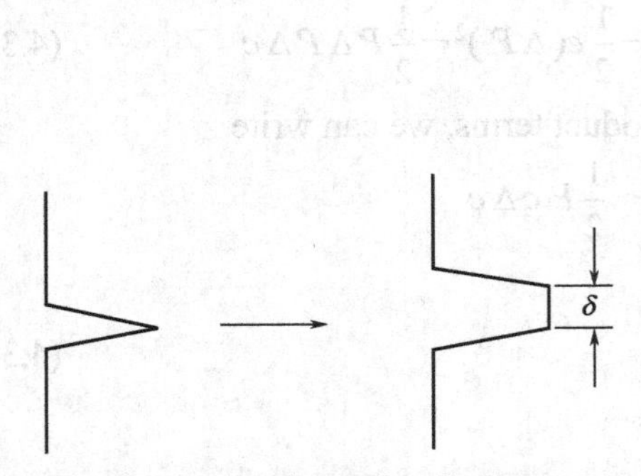

Figure 4.17 Crack opening displacement.

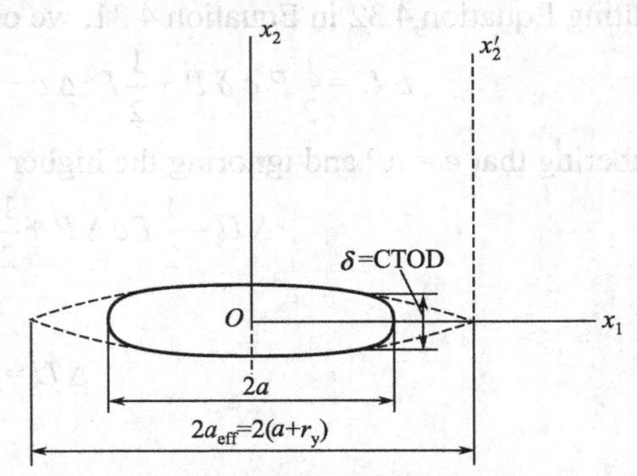

Figure 4.18 Relationship between crack opening displacement (COD, Δ), crack-tip opening displacement (CTOD, δ), crack length ($2a$), and size of plastic zone (r_y).

given for $x = a$ and $r_y << a$ (Figure 4.18) .Wells suggested that when this displacement at the crack tip reaches a critical value Δc, fracture would ensue.

LEFM is applicable only when the plastic zone is small in relation to the crack length (i.e., well below the yield stress and in plane strain). Consider a small crack in a brittle material. We have

$$\sigma_c = K_{Ic}(\sqrt{\pi a})^{-1}, \text{ as } a \to 0,\ \sigma_c \to \infty$$

But this, as we very well know, does not occur. Instead, a plastic zone develops and may extend through the section such that

$$\sigma_{net} = \sigma \frac{W}{W-a} \geqslant \sigma_y$$

where W is the width of sample and σ_y is the yield stress. In practice, $\sigma_c \leqslant 0.66\ \sigma_y$ for the K_{Ic} validity. In more ductile materials, the critical stress predicted by LEFM will be higher than σ_y.

The more usual way of obtaining δ is to use the equations of Dugdale—BCS model of the crack. According to Dugdale—BCS model (Bilby, Cotrell, Swinden, op. cit.; Dugdale, op. cit.)

$$\delta = \frac{8\sigma_y a}{\pi E} \ln \sec \frac{\pi a}{2\sigma_y}$$

Expanding the log sec function in series, we get

$$\delta = \frac{8\sigma_y a}{\pi E}\left[\frac{1}{2}\left(\frac{\pi a}{2\sigma_y}\right)^2 + \frac{1}{12}\left(\frac{\pi q}{2\sigma_y}\right)^4 + \frac{1}{45}\left(\frac{\pi a}{2\sigma_y}\right)^6 + \cdots\right]$$

For $\sigma << \sigma_y$, we can write (neglecting fourth and higher order terms)

$$\delta = \frac{\pi \sigma^2 a}{E\sigma_y} = \frac{G_I}{\sigma_y} \quad \text{(for plane stress)} \tag{4.36}$$

$$\delta = \frac{G_I}{nn\sigma_y} = \frac{\pi\sigma^2 a(1-\nu^2)}{nE\sigma_y} = \frac{K_I^2(1-\nu^2)}{nE\sigma_y} \quad \text{(for plane strain)} \tag{4.37}$$

The factor $(1-\nu^2)$ should be ignored in the case of plane stress. In the literature, we encounter various values of n. These depend on the exact location where CTOD is determined (i.e., the exact location of the crack tip). Wells suggested that, experimentally, $n \approx 2.1$ for compatibility with LEFM (i.e., limited plasticity). For cases involving extensive plasticity, the engineering design application approach is to take $n \approx 1$.

Thus, at unstable fracture, $G_{Ic} = n\ \sigma_y\ \delta_c$. The important point about COD is that, theoretically, δ_c can be computed for both elastic and plastic materials, whereas G_{Ic} is restricted only to the elastic regime.

4.6.3 *J* Integral

J integral is another variant for fracture toughness analysis. It provides a value of energy required to propagate a crack in an elastic—plastic material. The mathematical foundation for the *J* integral was laid by Eshelby, who applied it to dislocations. Cherepanov and Rice applied it, independently, to cracks. Figure 4.19 shows a closed contour Γ in a two-dimensional body. When such a body is subjected to external forces, internal stresses arise in it. On the basis of the theory of conservation of energy, Eshelby showed that the integral J is equal to zero for a closed contour; that is

$$J = \int_\Gamma \left(W \mathrm{d}x_2 - T\frac{\partial u}{\partial x_1}\right) = 0 \tag{4.38}$$

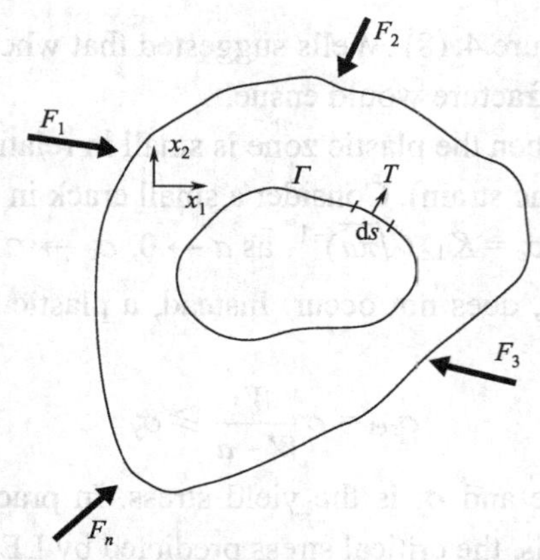

Figure 4.19 A body subjected to external forces F_1, F_2, …, F_n and with a closed contour Γ.

where $W = \int_0^{\Sigma_{ij}} \sigma_{ij} \mathrm{d}\varepsilon_{ij}$ is the strain energy per unit volume, T is the tension vector (traction) perpendicular to Γ and pointing to the outside of the contour, ds is an element of length along the contour, and u is the displacement in the x_1 direction. The J integral is an energy related quantity; similar to the crack extension force G, J has the units of energy per unit area (J/m^2) or force per unit length (N/m).

Figure 4.20 shows a crack, around which a contour $ABCDEFA$ is made. The total J must be zero, i.e.,

$$J = J_{\Gamma_1+\Gamma_2} = 0$$

Along AF and CD (crack surfaces), the tractions T are equal to zero.

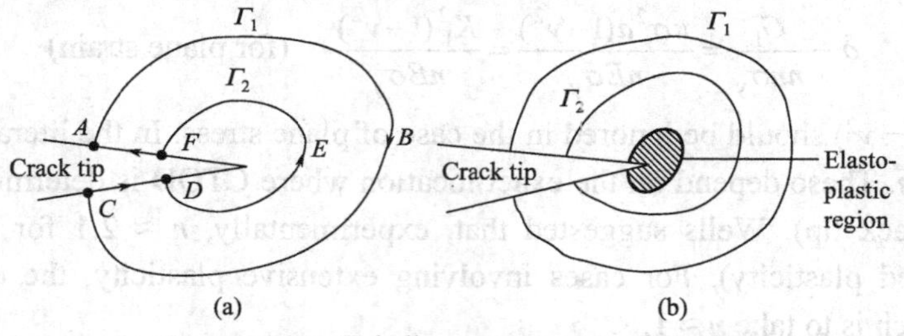

Figure 4.20 Eshelby contours around cracks.

The same is true for the normal and shear stresses. Thus, $J_{\mathrm{AF}} = J_{\mathrm{CD}} = 0$. It can therefore be concluded that

$$J_{\Gamma_1+\Gamma_2} = J_{\Gamma_1} + J_{\Gamma_2} + J_{\mathrm{AF}} + J_{\mathrm{CD}} = 0$$

$$J_{\Gamma_1} = -J_{\Gamma_2}$$

Hence, the J integral along two different paths around a crack has the same value. That is, in general, the J integral around a crack is path independent.

From a physical point of view, the J integral represents the difference in the potential energies of identical bodies containing cracks of length a and a + da; i n other words, the J integral around a crack is equal to the change in potential energy for a crack extension da.

For a body of thickness B, this can be written as

$$J = \frac{1}{B} \frac{\partial U}{\partial a} \tag{4.39}$$

where U is the potential energy, a the crack length, and B the plate thickness. U is equal to the area under the curve of load versus displacement. Figure 4.21 shows this interpretation, where the shaded area is $\Delta U = JB\Delta a$. Like G_{Ic}, J_{Ic} measures the critical energy associated with the initiation of crack growth, but in this case accompanied by substantial plastic deformation. In fact, Begley and Landes showed the formal equivalence of J_{Ic} and G_{Ic} by measuring the J_{Ic} from small fully plastic specimens and the G_{Ic} from large elastic specimens satisfying the plane-strain conditions for the LEFM test.

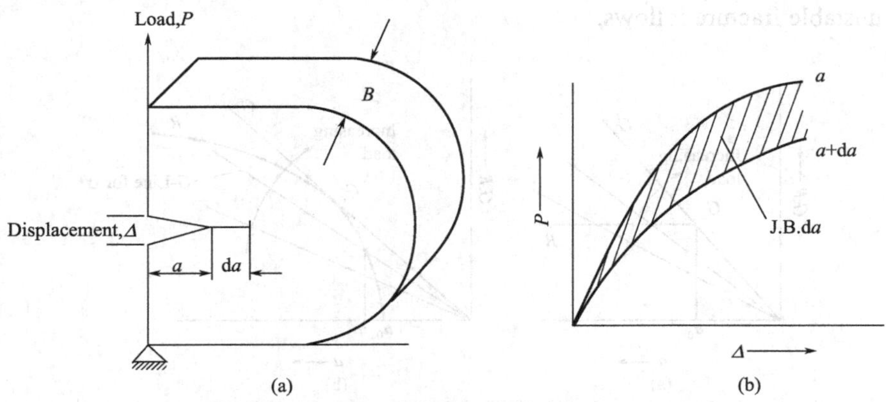

Figure 4.21 Physical interpretation of the J integral

The path independence of the J integral, together with this interpretation in terms of energy, makes it a powerful analytical tool. The J integral is path independent in the case of either linear or nonlinear materials behaving elastically. When extensive plastic deformation occurs, the practice is to assume that the plastic yielding can be described by the deformation theory of plasticity. According to this theory, stresses and strains are functions only of the point of measurement and not of the path taken to get to that point. As in the case of slow, stable crack growth, there will be a relaxation of stresses at the crack tip, so there will be a violation of this postulate. Thus, the use of the J integral should be limited to the initiation of crack propagation, by stable or unstable processes. Studies using incremental plasticity or flow theories with finite elements indicate the path independence of the J integral.

The J integral represents the difference in potential energy (shaded area) of identical bodies containing cracks of length a and $a + da$.

4.6.4 *R* Curve

The R curve characterizes the resistance of a material to fracture during slow and stable propagation of a crack. An R curve graphically represents this resistance to crack propagation of the material as a function of crack growth. With increasing load in a cracked structure, the crack extension force G at the crack tip also increases (Equation 4.34). However, the material at the tip presents a resistance R (sometimes, the symbol KR is used) to crack growth. According to Irwin, failure will occur when the rate of change of the crack extension force ($\Delta G/\Delta a$) equals the rate of change of this resistance to crack growth in the material ($\Delta R/\Delta a$). The resistance of the material to crack growth, R, increases with an increase in the size of the plastic zone. Since the plastic zone size increases nonlinearly with a, R will also be expected to increase nonlinearly with a. G increases linearly with a. Figure 4.22 shows the instability criterion: the point of tangency between the curves of G versus a and R versus a. Figure 4.22(a) shows the R curve for a brittle material, and Figure 4.22(b) shows the R curve for a ductile material. Crack extension occurs for $G > R$. Consider the G line for a stress σ, shown in Figure 4.22(b). At the stress σ, the crack in the material will grow only from a_0 to a, since $G > R$ for $a < a'$. $G < R$ for $a > a'$, and the crack does not extend beyond a. As the load is increased, the position of the G line changes, as indicated in the figure. When G becomes tangent to R, unstable fracture ensues. The R curve for a brittle material [Figure 4.22(a)] is a "square" curve, and the crack does not extend at all until the contact is reached, at which point $G = G_c$ and the unstable fracture follows.

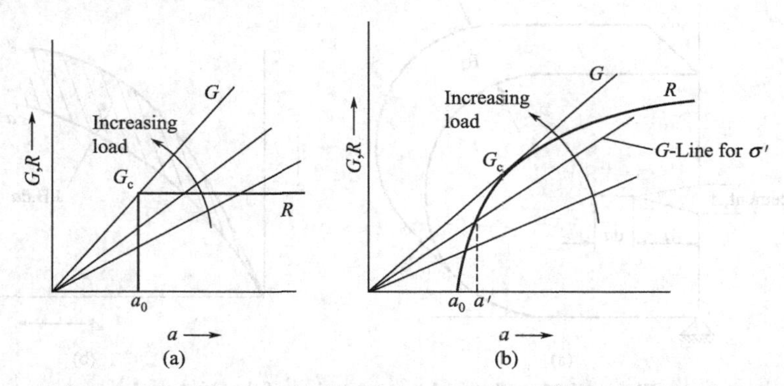

Figure 4.22 R curves for (a) brittle material and (b) ductile material.

The R curve method is another version of the Griffith energy balance. One can conveniently make this kind of analysis if an analytical expression for the R curve is available. Experimental determination of R curves, however, is complicated and time consuming.

4.6.5 Relationships among Different Fracture Toughness Parameters

So far, we have seen that, in our effort to develop a quantitative description of fracture toughness, various parameters, such as K, G, J, δ, R, etc., have been developed. Since all these parameters define the same physical quantity, it is not unexpected that they are interrelated. And we have mentioned in different sections the relationships among the parameters. Figure 4.23

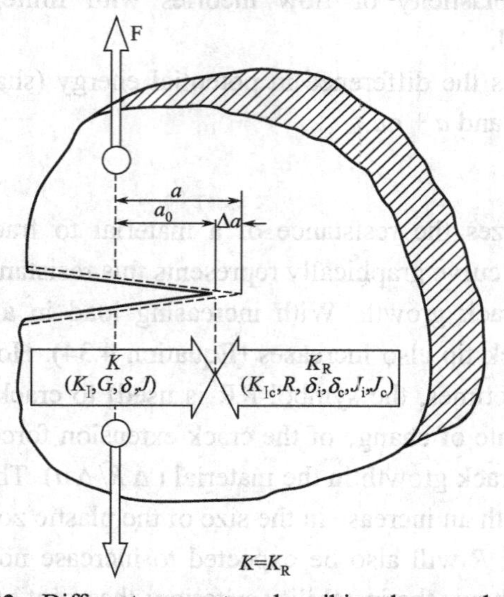

Figure 4.23 Different parameters describing the growth of a crack.

summarizes these relationships. It would, however, be helpful to the reader to recapitulate these relationships, even at the risk of repeating. That is what we will do in this section.

If we take into account the stress distribution around the tip of a crack, we get the stress-intensity-factor (K) approach. The magnitude or the intensity of the local stresses is determined by K, because the form of the local crack-tip stress field is the same for all situations involving a remote stress σ. Thus, K, and not σ, is the local characterizing parameter. The fracture then occurs when the applied K attains the critical value K_c. In particular, when the specimen's dimensions satisfy the plane-strain condition, we call this value the plane-strain fracture toughness and denote it by K_{Ic}. The stress and the crack length corresponding to K_{Ic} are the fracture stress σ_c and the fracture crack length a_c. Note that the elastic constants of the material are not involved. The energy-release-rate approach gives us the crack extension force G, which is related to the parameters K by the equation

$$K^2 = E'G,$$

where $E' = E$, Young's modulus, in the case of plane stress and $E' = E/(1 - \nu^2)$ in the case of plane strain. Note that, in characterizing the fracture behavior in terms of G, we need to know the elastic constants of the material. Because in the case of polymers E is time dependent and very precise modulus data are not available, there is some advantage to using the K approach.

The critical crack opening displacement Δc is another useful parameter. It is related to K by the equation $\delta_c = K_{Ic}^2/nE\sigma_y$, where n is a dimensionless constant that depends on the geometry of the specimen, its state of stress, and the work-hardening capacity of the material. λ has a value between 1 and 2. In particular, for the strip-yielding model of Dugdale—BCS, $n = 1$.

The J integral provides yet another measure of fracture toughness. And, for small-scale yielding, we have

$$J = n\delta\sigma_y$$

In short, for small-scale yielding, we can sum up the relationships among the different fracture toughness parameters as

$$J = G = K^2/E' = n\sigma_y\delta$$

where the symbols have the usual meaning.

EXAMPLE 4.2

Material Specification for a Pressurized Spherical Tank

Consider the thin-walled spherical tank of radius r and thickness t (Figure 4.24) that may be used as a pressure vessel.

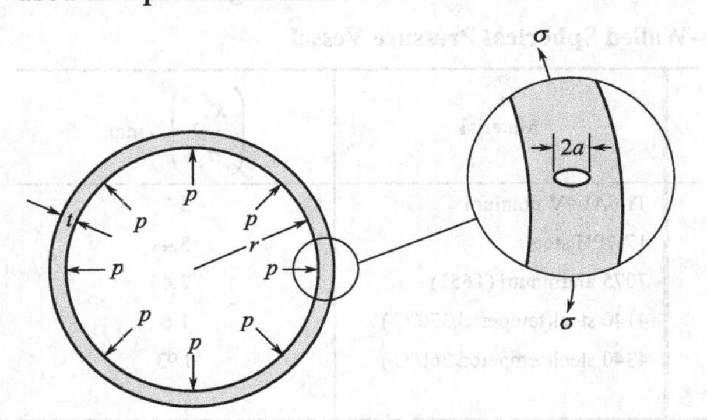

Figure 4.24 Schematic diagram showing the cross section of a spherical tank that is subjected to an internal pressure p, and that has a radial crack of length $2a$ in its wall.

(a) One design of such a tank calls for yielding of the wall material prior to failure as a result of the formation of a crack of critical size and its subsequent rapid propagation. Thus, plastic distortion of the wall may be observed and the pressure within the tank released before the occurrence of catastrophic failure. Consequently, materials having large critical crack lengths are desired. On the basis of this criterion, as to critical crack size, from longest to shortest.

(b) An alternative design that is also often utilized with pressure vessels is termed leak-before-break. Using principles of fracture mechanics, allowance is made for the growth of a crack through the thickness of the vessel wall prior to the occurrence of rapid crack propagation (Figure 4.24).Thus, the crack will completely penetrate the wall without catastrophic failure, allowing for its detection by the leaking of pressurized fluid. With this criterion the critical crack length α_c (i.e., one-half of the total internal crack length) is taken to be equal to the pressure vessel thickness t. Allowance for $\alpha_c = t$ instead of $\alpha_c = t/2$ assures that fluid leakage will occur prior to the buildup of dangerously high pressures.

For this spherical pressure vessel, the circumferential wall stress σ is a function of the pressure p in the vessel and the radius r and wall thickness t according to

$$\sigma = \frac{pr}{2t} \tag{4.40}$$

For both parts (a) and (b) assume a condition of plane strain.

Solution

(a) For the first design criterion, it is desired that the circumferential wall stress be less than the yield strength of the material. Substitution of σ_y for σ in Equation 3.5, and incorporation of a factor of safety N leads to

$$K_{\mathrm{I\,c}} = Y\left(\frac{\sigma_y}{N}\right)\sqrt{\pi a_c} \tag{4.41}$$

where a_c is the critical crack length. Solving for a_c yields the following expression:

$$a_c = \frac{N^2}{Y^2\pi}\left(\frac{K_{\mathrm{I\,c}}}{\sigma_y}\right)^2 \tag{4.42}$$

Therefore, the critical crack length is proportional to the square of the $K_{\mathrm{I\,c}}$-σ_y ratio. The ranking is provided in Table 4.3, where it may be seen that the medium carbon (1040) steel with the largest ratio has the longest critical crack length, and, therefore, is the most desirable material on the basis of this criterion.

Table 4.3 Ranking of Several Metal Alloys Relative to Critical Crack Length (Yielding Criterion) for a Thin-Walled Spherical Pressure Vessel

Material	$\left(\frac{K_{\mathrm{I\,c}}}{\sigma_y}\right)^2$/mm	Material	$\left(\frac{K_{\mathrm{I\,c}}}{\sigma_y}\right)^2$/mm
Medium carbon (1040) steel	43.1	Ti-6Al-4V titanium	3.7
AZ31B magnesium	19.6	17-7PH steel	3.4
2024 aluminum (T3)	16.3	7075 aluminum (T651)	2.4
Ti-5Al-2.5Sn titanium	6.6	4140 steel(tempered 370℃)	1.6
4140 steel(tempered 482℃)	5.3	4340 steel(tempered 260℃)	0.93
4340 steel(tempered 425℃)	3.8		

(b) As stated previously, the leak-before-break criterion is just met when one half of the internal crack length is equal to the thickness of the pressure vessel—that is, when $a=t$. Substitution of $a=t$ into Equation 3.5 gives

$$K_{\mathrm{Ic}}=Y\sigma\sqrt{\pi t} \tag{4.43}$$

and, from Equation 3.8

$$t=\frac{pr}{2\sigma} \tag{4.44}$$

The stress is replaced by the yield strength, inasmuch as the tank should be designed to contain the pressure without yielding; furthermore, substitution of Equation 3.12 into Equation 3.11, after some rearrangement, yields the following expression:

$$p=\frac{2}{Y^2\pi r}\left(\frac{K_{\mathrm{Ic}}^2}{\sigma_{\mathrm{y}}}\right) \tag{4.45}$$

Hence, for some given spherical vessel of radius r, the maximum allowable pressure consistent with this leak-before-break criterion is proportional to $K_{\mathrm{Ic}}^2/\sigma_{\mathrm{y}}$. The same several materials are ranked according to this ratio in Table 4.4; as may be noted, the medium carbon steel will contain the greatest pressures.

Table 4.4 Ranking of Several Metal Alloys Relative to Maximum Allowable Pressure (Leak-Before-Break Criterion) **for a Thin-Walled Spherical Pressure Vessel**

Material	$\frac{K_{\mathrm{Ic}}^2}{\sigma_{\mathrm{y}}}$ /(MPa • m)	Material	$\frac{K_{\mathrm{Ic}}^2}{\sigma_{\mathrm{y}}}$ /(MPa • m)
Medium carbon (1040) steel	11.2	AZ31B magnesium	3.9
4140 steel(tempered 482℃)	6.1	Ti-6Al-4V titanium	3.3
Ti-5Al-2.5Sn titanium	5.8	4140 steel(tempered 370℃)	2.4
2024 aluminum (T3)	5.6	4340 steel(tempered 260℃)	1.5
4340 steel(tempered 425℃)	5.4	7075 aluminum (T651)	1.2
17-7PH steel	4.4		

4.7 Impact Fracture

Prior to the advent of fracture mechanics as a scientific discipline, impact testing techniques were established so as to ascertain the fracture characteristics of materials. It was realized that the results of laboratory tensile tests could not be extrapolated to predict fracture behavior; for example, under some circumstances normally ductile metals fracture abruptly and with very little plastic deformation. Impact test conditions were chosen to represent those most severe relative to the potential for fracture—namely, (1) deformation at a relatively low temperature, (2) a high strain rate (i.e., rate of deformation), and (3) a triaxial stress state (which may be introduced by the presence of a notch).

4.7.1 Impact Testing Techniques

Two standardized tests (ASTM Standard E 23, “Standard Test Methods for Notched Bar Impact Testing of Metallic Materials.”), the Charpy and Izod, were designed and are still used to measure the impact energy, sometimes also termed notch toughness. The Charpy V-notch (CVN) technique is most commonly used in the United States. For both Charpy and Izod, the specimen is in the shape of a bar of square cross section, into which a V-notch is machined [Figure 4.25(a)]. The apparatus for making V-notch impact tests is illustrated schematically in Figure 4.25(b). The load is applied as an impact blow from a weighted pendulum hammer that is released from a cocked position at a fixed height *h*. The specimen is positioned at the base as shown.

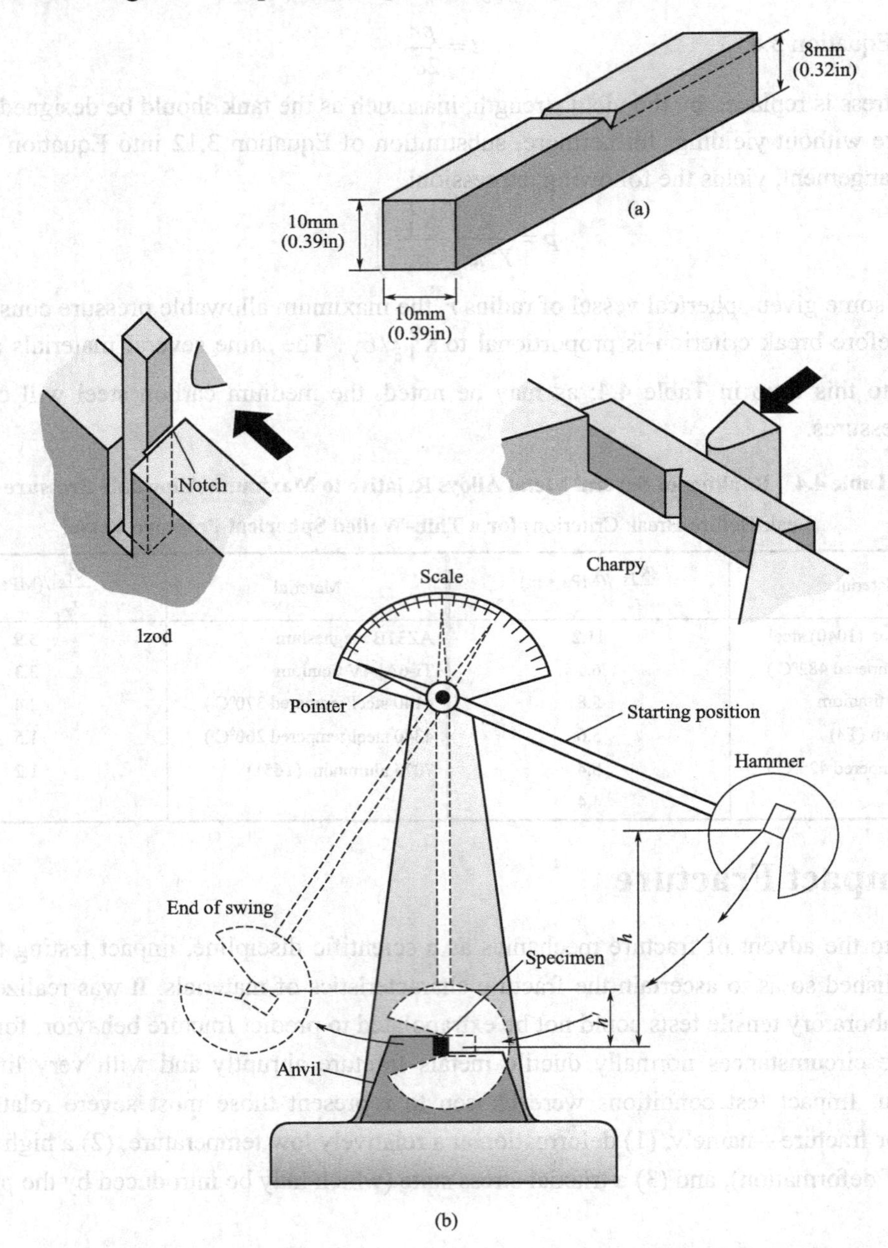

Figure 4.25 (a) Specimen used for Charpy and Izod impact tests. (b) A schematic drawing of an impact testing apparatus. The hammer is released from fixed height *h* and strikes the specimen; the energy expended in fracture is reflected in the difference between *h* and the swing height Specimen placements for both Charpy and Izod tests are also shown. [Figure (b) adapted from H.W. Hayden, W. G. Moffatt, and J.Wulff, *The Structure and Properties Materials,* Vol. III, Mechanical Behavior, p. 13. Copyright © 1965 by John Wiley & Sons, New York. Reprinted by permission of John Wiley & Sons, Inc.]

Upon release, a knife edge mounted on the pendulum strikes and fractures the specimen at the notch, which acts as a point of stress concentration for this high-velocity impact blow. The pendulum continues its swing, rising to a maximum height h',which is lower than h. The energy absorption, computed from the difference between h and $h'$is a measure of the impact energy. The primary difference between the Charpy and Izod techniques lies in the manner of specimen support, as illustrated in Figure 4.25(b). Furthermore, these are termed impact tests in light of the manner of load application. Variables including specimen size and shape as well as notch configuration and depth influence the test results.

Both plane strain fracture toughness and these impact tests determine the fracture properties of materials. The former are quantitative in nature, in that a specific property of the material is determined (i.e., K_{Ic}). The results of the impact tests, on the other hand, are more qualitative and are of little use for design purposes. Impact energies are of interest mainly in a relative sense and for making comparisons—absolute values are of little significance. Attempts have been made to correlate plane strain fracture toughnesses and CVN energies, with only limited success. Plane strain fracture toughness tests are not as simple to perform as impact tests; furthermore, equipment and specimens are more expensive.

4.7.2 Ductile-to-Brittle Transition

One of the primary functions of Charpy and Izod tests is to determine whether or not a material experiences a ductile-to-brittle transition with decreasing temperature and, if so, the range of temperatures over which it occurs. The ductile-to-brittle transition is related to the temperature dependence of the measured impact energy absorption. This transition is represented for a steel by curve *A* in Figure 4.26. At higher temperatures the CVN energy is relatively large, in correlation with a ductile mode of fracture. As the temperature is lowered, the impact energy drops suddenly over a relatively narrow temperature range, below which the energy has a constant but small value; that is, the mode of fracture is brittle.

Alternatively, appearance of the failure surface is indicative of the nature of fracture and may be used in transition temperature determinations. For ductile fracture this surface appears fibrous or dull (or of shear character), as in the steel specimen of Figure 4.27 that was tested at 79℃. Conversely,

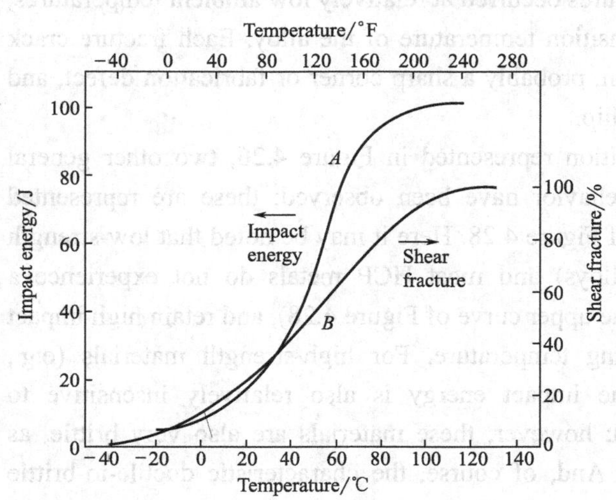

Figure 4.26 Temperature dependence of the Charpy V-notch impact energy (curve *A*) and percent shear fracture (curve *B*) for an A283 steel. (Reprinted from Welding Journal. Used by permission of the American Welding Society.)

Figure 4.27 Photograph of fracture surfaces of A36 steel Charpy V-notch specimens tested at indicated temperatures (in ºC). (From R.W. Hertzberg, Deformation and Fracture Mechanics of Engineering Materials, 3rd edition, Fig. 9.6, p. 329. Copyright © 1989 by John Wiley & Sons, Inc., New York. Reprinted by permission of John Wiley & Sons, Inc.)

totally brittle surfaces have a granular (shiny) texture (or cleavage character) (the −59ºC specimen, Figure 4.27). Over the ductile-to-brittle transition, features of both types will exist (in Figure 4.27, displayed by specimens tested at −12ºC, 4ºC, 16ºC and 24ºC). Frequently, the percent shear fracture is plotted as a function of temperature—curve *B* in Figure 4.26.

For many alloys there is a range of temperatures over which the ductile-to-brittle transition occurs (Figure 4.26); this presents some difficulty in specifying a single ductile-to-brittle transition temperature. No explicit criterion has been established, and so this temperature is often defined as that temperature at which the CVN energy assumes some value (e.g., 20 J or 15 ft-lbf), or corresponding to some given fracture appearance (e.g., 50% fibrous fracture). Matters are further complicated in as much as a different transition temperature may be realized for each of these criteria. Perhaps the most conservative transition temperature is that at which the fracture surface becomes 100% fibrous; on this basis, the transition temperature is approximately 110ºC (230ºF) for the steel alloy that is the subject of Figure 4.26.

Structures constructed from alloys that exhibit this ductile-to-brittle behavior should be used only at temperatures above the transition temperature, to avoid brittle and catastrophic failure. Classic examples of this type of failure occurred, with disastrous consequences, during World War II when a number of welded transport ships, away from combat, suddenly and precipitously split in half. The vessels were constructed of a steel alloy that possessed adequate ductility according to room-temperature tensile tests. The brittle fractures occurred at relatively low ambient temperatures, at about 4ºC (40ºF), in the vicinity of the transition temperature of the alloy. Each fracture crack originated at some point of stress concentration, probably a sharp corner or fabrication defect, and then propagated around the entire girth of the ship.

In addition to the ductile-to-brittle transition represented in Figure 4.26, two other general types of impact energy-versus-temperature behavior have been observed; these are represented schematically by the upper and lower curves of Figure 4.28. Here it may be noted that low-strength FCC metals (some aluminum and copper alloys) and most HCP metals do not experience a ductile-to-brittle transition (corresponding to the upper curve of Figure 4.28), and retain high impact energies (i.e., remain ductile) with decreasing temperature. For high-strength materials (e.g., high-strength steels and titanium alloys), the impact energy is also relatively insensitive to temperature (the lower curve of Figure 4.28); however, these materials are also very brittle, as reflected by their low impact energy values. And, of course, the characteristic ductile-to-brittle transition is represented by the middle curve of Figure 4.28.

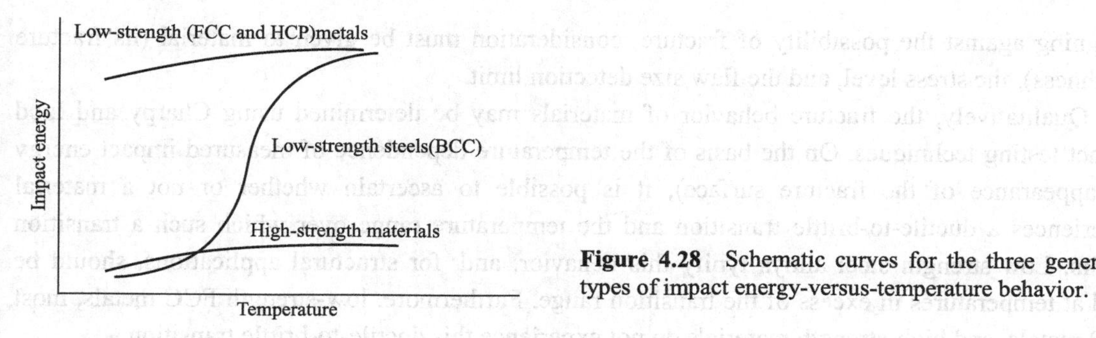

Figure 4.28 Schematic curves for the three general types of impact energy-versus-temperature behavior.

As noted, this behavior is typically found in low-strength steels that have the BCC crystal structure.

For these low-strength steels, the transition temperature is sensitive to both alloy composition and microstructure. For example, decreasing the average grain size results in the lowering of the transition temperature. Hence, refining the grain size both strengthens and toughens steels. In contrast, increasing the carbon content, while increasing the strength of steels, also raises the CVN transition of steels, as indicated in Figure 4.29.

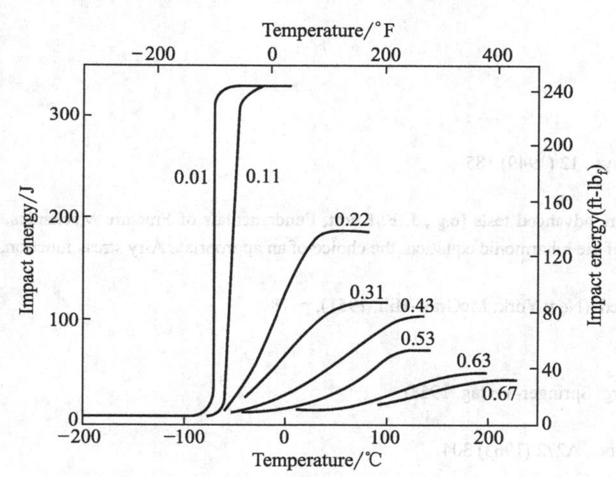

Figure 4.29 Influence of carbon content on the Charpy V-notch energy –versus temperature behavior for steel. (Reprinted with permission from ASM International, Metals Park, OH 44073-9989, USA; J. A. Reinbolt and W. J. Harris, Jr., "Effect of Alloying Elements on Notch Toughness of Pearlitic Steels," *Transactions of ASM,* Vol. 43, 1951.)

Most ceramics and polymers also experience a ductile-to-brittle transition. For ceramic materials, the transition occurs only at elevated temperatures, ordinarily in excess of 1000ºC (1850ºF).

SUMMARY

The significant discrepancy between actual and theoretical fracture strengths of brittle materials is explained by the existence of small flaws that are capable of amplifying an applied tensile stress in their vicinity, leading ultimately to crack formation. Fracture ensues when the theoretical cohesive strength is exceeded at the tip of one of these flaws.

The fracture toughness of a material is indicative of its resistance to brittle fracture when a crack is present. It depends on specimen thickness, and, for relatively thick specimens (i.e., conditions of plane strain), is termed the plane strain fracture toughness. This parameter is the one normally cited for design purposes; its value is relatively large for ductile materials (and small for brittle ones), and is a function of microstructure, strain rate, and temperature. With regard to designing against the possibility of fracture, consideration must be given to material (its fracture toughness), the stress level, and the flaw size detection limit.

Qualitatively, the fracture behavior of materials may be determined using Charpy and Izod impact testing techniques. On the basis of the temperature dependence of measured impact energy (or appearance of the fracture surface), it is possible to ascertain whether or not a material experiences a ductile-to-brittle transition and the temperature range over which such a transition occurs. Low-strength steel alloys typify this behavior, and, for structural applications, should be used at temperatures in excess of the transition range. Furthermore, low-strength FCC metals, most HCP metals, and high-strength materials do not experience this ductile-to-brittle transition.

IMPORTANT TERMS AND CONCEPTS

Linear elastic fracture mechanics (LEFM)
Theoretical cleavage strength
Stress concentration
Stress concentration Factor
Fracture toughness
Charpy test
Ductile-to-brittle transition

REFERENCES

1. E. Orowan, "Fracture and Strength of Solids," Rep. Prog. Phys., 12 (1949) 185
2. C. E. Inglis, Proc. Inst. Naval Arch., 55 (1913) 163, 219
3. The derivation of this equation, which can be found in more advanced tests [e.g., J. F. Knott, Fundamentals of Fracture Mechanics, (London: Butterworths, 1973), p. 51], involves the solution of the biharmonic equation, the choice of an appropriate Airy stress function, and complex variables
4. S. Timoshenko and J. N. Goodier, Theory of Elasticity, 2nd ed. (New York: McGraw-Hill, 1951), p. 78
5. J. N. Goodier, App. Mech. 1 (1933) 39
6. H. M. Westergaard, J. Appl. Mechan., 5A (1939) 49
7. G. R. Irwin, in Encyclopaedia of Physics, Vol. VI (Heidelberg: Springer-Verlag, 1958)
8. G. R Irwin, J. Basic Eng., Trans. ASME, 82 (1960) 417
9. B. A. Bilby, A. H. Cottrell, and K. H. Swinden, Proc. Roy. Soc., A272 (1963) 304
10. D. S. Dugdale, J. Mech. Phys. Solids, 8 (1960) 100
11. A. A. Wells, Brit. Weld. J., 13 (1965) 2
12. A A. Wells, Eng. Fract. Mech., 1 (1970) 399
13. J. D. Eshelby, Phil. Trans. Roy. Soc LondonN, A244 (1951) 87
14. G. P. Cherepanov, Appl. Math. Mech. (Prinkl. Mat. Mekh.), 31, no. 3 (1967) 503
15. J. R. Rice, J. Appl. Mech., 35 (1968) 379
16 J. A. Begley and J. D. Landes, ASTM STP 514, (Philadelphia: ASTM, 1972), p. 1
17. T. L. Anderson. Fracture Mechanics, 2nd ed. Boca Raton, Fl: CRC Press, 1995
18. J. M. Barsom and S. T. Roffe. Fracture and Fatigue Control in Structures, 2nd ed. Englewood Cliffs, NJ: Prentice Hall, 1987
19. D. Broek. Elementary Engineering Fracture Mechanics, 3rd ed. The Hague: Sijthoff and Noordhoff, 1978
20. H. L. Ewalds and R. J. H. Wanhill. Fracture Mechanics. London: Arnold, 1984
21. M. F. Kanninen and C. H. Popelar. Advanced Fracture Mechanics. NewYork: Oxford University Press, 1985
22. J. F. Knott. Fundamentals of Fracture Mechanics, 3rd ed. London: Butterworths, 1993
23. R. J. Sanford, Principles of Fracture Mechanics. Upper Saddle River, NJ: Prentice Hall, 2003
24. H. Tada, P. C. Paris, and G. R. Irwin, The Stress Analysis of Cracks Handbook, 3rd ed., New York: ASME, 2000

QUESTIONS AND PROBLEMS

4.1 What is the magnitude of the maximum stress that exists at the tip of an internal crack having a radius of curvature of 1.9×10^{-4}mm (7.5×10^{-6} in) and a crack length of 3.8×10^{-2} mm (1.5×10^{-3} in) when a tensile stress of 140 MPa (20000 psi) is applied?

4.2 Estimate the theoretical fracture strength of a brittle material if it is known that fracture occurs by the propagation of an elliptically shaped surface crack of length 0.5 mm (0.02 in) and having a tip radius of curvature of 5×10^{-3} mm (2×10^{-4} in), when a stress of 1035 MPa (150000 psi) is applied.

4.3 If the specific surface energy for aluminum oxide is 0.90 J/m², using data contained in Table 4.1, compute the critical stress required for the propagation of an internal crack of length 0.40 mm.

4.4 A MgO component must not fail when a tensile stress of 13.5 MPa (1960 psi) is applied. Determine the maximum allowable surface crack length if the surface energy of MgO is 1.0 J/m². Data found in Table 4.1 may prove helpful.

4.5 A specimen of a 4340 steel alloy with a plane strain fracture toughness of 54.8 MPa $\sqrt{m}$ (50 ksi $\sqrt{in}$) is exposed to a stress of 1030 MPa (150000 psi). Will this specimen experience fracture if it is known that the largest surface crack is 0.5 mm (0.02 in) long? Why or why not? Assume that the parameter Y has a value of 1.0.

4.6 Some aircraft component is fabricated from an aluminum alloy that has a plane strain fracture toughness of 40 MPa $\sqrt{m}$ (36.4 ksi $\sqrt{in}$). It has been determined that fracture results at a stress of 300 MPa (43500 psi) when the maximum (or critical) internal crack length is 4.0 mm (0.16 in). For this same component and alloy, will fracture occur at a stress level of 260 MPa (38000 psi) when the maximum internal crack length is 6.0 mm (0.24 in)? Why or why not?

4.7 Suppose that a wing component on an aircraft is fabricated from an aluminum alloy that has a plane strain fracture toughness of 26 MPa $\sqrt{m}$ (23.7 ksi $\sqrt{in}$). It has been determined that fracture results at a stress of 112 MPa (16240 psi) when the maximum internal crack length is 3.6 mm (0.34 in). For this same component and alloy, compute the stress level at which fracture will occur for a critical internal crack length of 6.0 mm (0.24 in).

4.8 A large plate is fabricated from a steel alloy that has a plane strain fracture toughness of 82.4 MPa $\sqrt{m}$ (75.0 ksi $\sqrt{in}$). If, during service use, the plate is exposed to a tensile stress of 345 MPa (50000 psi), determine the minimum length of a surface crack that will lead to fracture. Assume a value of 1.0 for Y.

4.9 Calculate the maximum internal crack length allowable for a Ti-6Al-4V titanium alloy (Table 3.1) component that is loaded to a stress one-half of its yield strength. Assume that the value of Y is 1.50.

4.10 A structural component in the form of a wide plate is to be fabricated from a steel alloy that has a plane strain fracture toughness of and a yield strength of 860 MPa (125000 psi). The flaw size resolution limit of the flaw detection apparatus is 3.0 mm (0.12 in). If the design stress is one half of the yield strength and the value of Y is 1.0, determine whether or not a critical flaw for this plate is subject to detection.

4.11 After consultation of other references, write a brief report on one or two nondestructive test techniques that are used to detect and measure internal and/or surface flaws in metal alloys.

4.12 Following is tabulated data that were gathered from a series of Charpy impact tests on a tempered 4340 steel alloy.

Temperature/°C	Impact Energy/J	Temperature/°C	Impact Energy/J
0	105	−113	40
−25	104	−125	34
−50	103	−150	28
−75	97	−175	25
−100	63	−200	24

(a) Plot the data as impact energy versus temperature.

(b) Determine a ductile-to-brittle transition temperature as that temperature corresponding to the average of the maximum and minimum impact energies.

(c) Determine a ductile-to-brittle transition temperature as that temperature at which the impact energy is 50 J.

4.13 Calculate the maximum tensile stress at the surfaces of a circular hole (in the case of a thin sheet) and of a spherical hole (in the case of a thick specimen) subjected to a tensile stress of 200 MPa. The material is Al_2O_3 with $\nu = 0.2$.

4.14 Calculate the maximum tensile stress if the applied stress is compressive for a circular hole for which σ_c = 200 MPa and $\nu = 0.2$.

4.15 The strength of alumina is approximately $E/15$, where E is the Young's modulus of alumina, equal to 380 GPa. Use the Griffith equation in the plane strain form to estimate the critical size of defect corresponding to fracture of alumina.

4.16 Compute the ratio of stress required to propagate a crack in a brittle material under plane-stress and plane-strain conditions. Take Poisson's ratio ν of the material to be 0.3.

4.17 An Al_2O_3 specimen is being pulled in tension. The specimen contains flaws having a size of 100 μm. If the surface energy of Al_2O_3 is 0.8 J/m^2, what is the fracture stress? Use Griffith's criterion. E = 380 GPa.

4.18 A central through-the-thickness crack, 50 mm long, propagates in a thermoset polymer in an unstable manner at an applied stress of 5 MPa. Find K_c.

4.19 Machining of SiC produced surface flaws of a semielliptical geometry. The flaws that were generated have dimensions a = 1 mm, width w = 100 mm, and c = 5 mm, and the thickness of the specimen is B = 20 mm. Calculate the maximum stress that the specimen can withstand in tension. K_{Ic} = 4 MPa m$^{1/2}$.

4.20 (a) An AISI 4340 steel plate has a width W of 30 cm and has a central crack $2a$ of 3 mm. The plate is under a uniform stress σ. This steel has a K_{Ic} value of 50 MPa m$^{1/2}$. Find the maximum stress for this crack length.

(b) If the operating stress is 1500 MPa, compute the maximum crack size that the steel may have without failure.

4.21 A microalloyed steel, quenched and tempered at 250°C, has a yield strength (σ_y) of 1750 MPa and a plane-strain fracture toughness K_{Ic} of 43.50 MPa m$^{1/2}$. What is the largest disk-type inclusion, oriented most unfavorably, that can be tolerated in this steel at an applied stress of 0.5 σ_y?

4.22 A 25mm^2 bar of cast iron contains a crack 5 mm long and normal to one face. What is the load required to break this bar if it is subjected to three point bending with the crack toward the tensile side and the supports 250 mm apart?

4.23 Consider a maraging steel plate of thickness (B) 3 mm. Two specimens of width (W) equal to 50 mm and 5 mm were taken out of this plate. What is the largest through-the-thickness crack that can be tolerated in the two cases at an applied stress of σ= 0.6 σ_y, where σ_y (yield stress) =2.5 GPa? The plane-strain fracture toughness K_{Ic} of the steel i s 70 MPa m$^{1/2}$. What are the critical dimensions in the case of a single-edge notch specimen?

4.24 An infinitely large plate containing a central crack of length $2a = 50/\pi$ mm is subjected to a nominal stress of 300 MPa. The material yields at 500 MPa. Compute:

(a) The stress intensity factor at the crack tip.

(b) The size of the plastic zone at the crack tip.

Comment on the validity of Irwin's correction for the size of the plastic zone in this case.

4.25 A steel plate containing a through-the-thickness central crack of length 15 mm is subjected to a stress of 350 MPa normal to the crack plane. The yield stress of the steel is 1500 MPa. Compute the size of the plastic zone and the effective stress intensity factor.

4.26 A sheet of polystyrene has a thin central crack with $2a$ = 50 mm. The crack propagates catastrophically at an applied stress of 10 MPa. The Young's modulus polystyrene is 3.8 GPa, and the Poisson's ratio is 0.4. Find G_{Ic}.

4.27 Compute the approximate size of the plastic zone, r_y , for an alloy that has a Young's modulus E = 70 GPa, yield strength σ_y = 500 MPa, and toughness G_c = 20 kJ/m^2.

4.28 300-M steel, commonly used for airplane landing gears, has a G_c value of 10 kN/m. A nondestructive examination technique capable of detecting cracks that are 1 mm long is available. Compute the stress level that the landing gear can support without failure.

4.29 A thermoplastic material has a yield stress of 75 MPa and a G_{Ic} value of 300 J/m^2. What would be the corresponding critical crack opening displacement? Take n = 1. Also, compute J_{Ic}.

4.30 A line pipe with overall diameter of 1 m and 25-mm thickness is constructed from a microalloyed steel (K_{Ic} =

60 MPa m$^{1/2}$; σ_y = 600 MPa). Calculate the maximum pressure for which the leak-before-break criterion will be obeyed. The leak-before-break criterion states that a through-the-thickness crack ($a = t$) will not propagate catastrophically.

Chapter 4

IMPORTANT TERMS AND CONCEPTS

Fracture Mechanics	断裂力学	Plastic-Zone Size	塑性区宽度
Linear Elastic Fracture Mechanics (LEFM)	线弹性断裂力学	Opening Mode	张开型（Ⅰ）
Fracture Toughness	断裂韧度	Sliding Mode	滑开型（Ⅱ）
Theoretical Cleavage Strength	理论断裂强度	Tearing Mode	撕开型（Ⅲ）
Orowan'S Theory	奥罗万理论	Crack Extension Force	裂纹扩展力
Stress Concentration Factor, K_t	应力集中系数	Energy Release Rate	能量释放率
Stress Intensity Factor, K	应力场强度因子	Crack Extension Energy Release Rate	裂纹能量释放率
Griffith Criterion of Fracture	格雷菲斯断裂判据	Crack Tip Opening Displacement (CTOD)	裂纹尖端张开位移
Plane Strain	平面应变	Crack Opening Displacement (COD)	裂纹张开位移
Plane Stress	平面应力	J Integral	J积分
Normal Stress	正应力		
Shear Stress	切应力		

Chapter 5

Fatigue of metals

Learning Objectives

After studying this chapter you should be able to do the following:

1. Define fatigue and specify the conditions under which it occurs.

2. From a fatigue plot for some material, determine (a) the fatigue lifetime (at a pecified stress level), and (b) the fatigue strength (at a specified number of cycles).

5.1 Introduction

Fatigue is a form of failure that occurs in structures subjected to dynamic and fluctuating stresses (e.g., bridges, aircraft, and machine components). Under these circumstances it is possible for failure to occur at a stress level considerably lower than the tensile or yield strength for a static load. The term "fatigue" is used because this type of failure normally occurs after a lengthy period of repeated stress or strain cycling. Fatigue is important inasmuch as it is the single largest cause of failure in metals, estimated to comprise approximately 90% of all metallic failures; polymers and ceramics (except for glasses) are also susceptible to this type of failure.

Furthermore, fatigue is catastrophic and insidious, occurring very suddenly and without warning. Fatigue failure is brittlelike in nature even in normally ductile metals, in that there is very little, if any, gross plastic deformation associated with failure. The process occurs by the initiation and propagation of cracks, and ordinarily the fracture surface is perpendicular to the direction of an applied tensile stress.

5.2 Cyclic Stresses

The applied stress may be axial (tension-compression), flexural (bending), or torsional (twisting) in nature. In general, three different fluctuating stress–time modes are possible. One is represented schematically by a regular and sinusoidal time dependence in Figure 5.1(a), wherein the amplitude is symmetrical about a mean zero stress level, for example, alternating from a maximum tensile stress (σ_{max}) to a minimum compressive stress (σ_{min}) of equal magnitude; this is referred to as a reversed stress cycle.

Also indicated in Figure 5.1(b) are several parameters used to characterize the fluctuating stress cycle. The stress amplitude alternates about a mean stress defined as the average of the maximum and minimum stresses in the cycle, or

$$\sigma_m = \frac{\sigma_{max} + \sigma_{min}}{2} \tag{5.1}$$

Furthermore, the range of stress is just the difference between σ_{max} and σ_{min}— namely,

$$\sigma_r = \sigma_{max} - \sigma_{min} \tag{5.2}$$

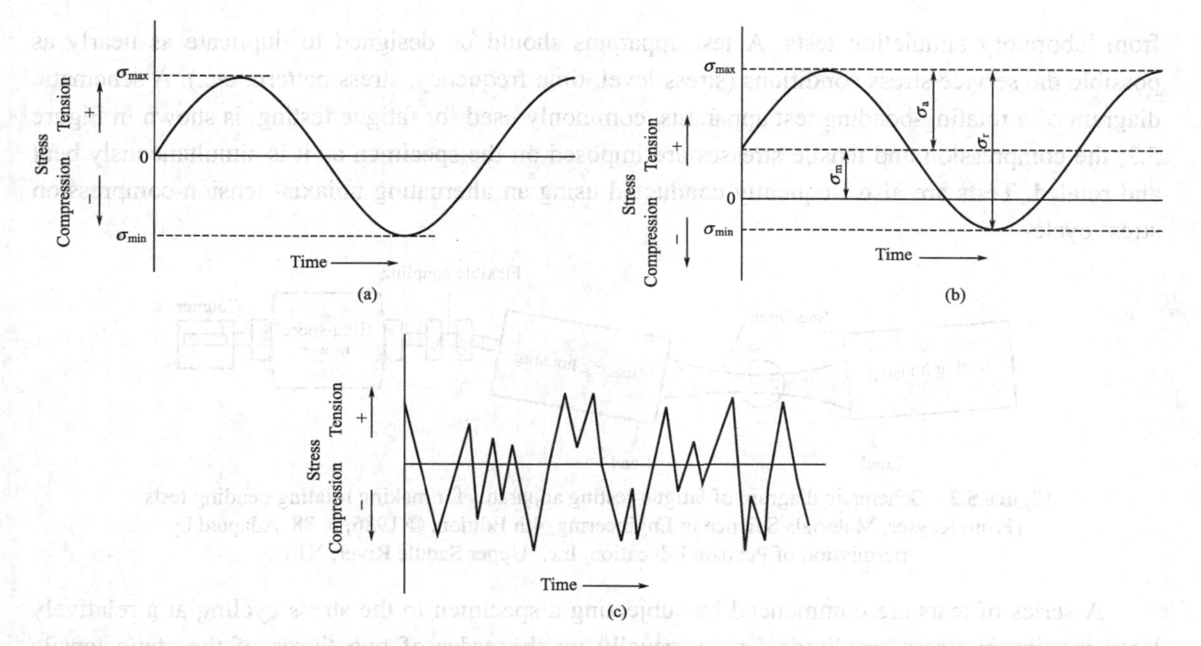

Figure 5.1 Variation of stress with time that accounts for fatigue failures. (a) Reversed stress cycle, in which the stress alternates from a maximum tensile stress (+) to a maximum compressive stress of equal magnitude. (b) Repeated stress cycle, in which maximum and minimum stresses are asymmetrical relative to the zero-stress level; mean stress σ_m, range of stress σ_r, and stress amplitude σ_a are indicated. (c) Random stress cycle.

Stress amplitude σ_a is just one half of this range of stress, or

$$\sigma_a = \sigma_r = \frac{\sigma_{max} - \sigma_{min}}{2} \tag{5.3}$$

Another type, termed repeated stress cycle, is illustrated in Figure 5.1(b); the maxima and minima are asymmetrical relative to the zero stress level. Finally, the stress level may vary randomly in amplitude and frequency, as exemplified in Figure 5.1(c).

Finally, the stress ratio R is just the ratio of minimum and maximum stress amplitudes:

$$R = \frac{\sigma_{min}}{\sigma_{max}} \tag{5.4}$$

By convention, tensile stresses are positive and compressive stresses are negative. For example, for the reversed stress cycle, the value of R is -1.

Concept Check 5.1

Make a schematic sketch of a stress-versus-time plot for the situation when the stress ratio R has a value of $+1$.

Concept Check 5.2

Using Equation 5.3 and Equation 5.4, demonstrate that increasing the value of the stress ratio R produces a decrease in stress amplitude σ_a.

5.3 The *S-N* Curve

As with other mechanical characteristics, the fatigue properties of materials can be determined from laboratory simulation tests. A test apparatus should be designed to duplicate as nearly as possible the service stress conditions (stress level, time frequency, stress pattern, etc.). A schematic diagram of a rotating-bending test apparatus, commonly used for fatigue testing, is shown in Figure 5.2; the compression and tensile stresses are imposed on the specimen as it is simultaneously bent and rotated. Tests are also frequently conducted using an alternating uniaxial tension-compression stress cycle.

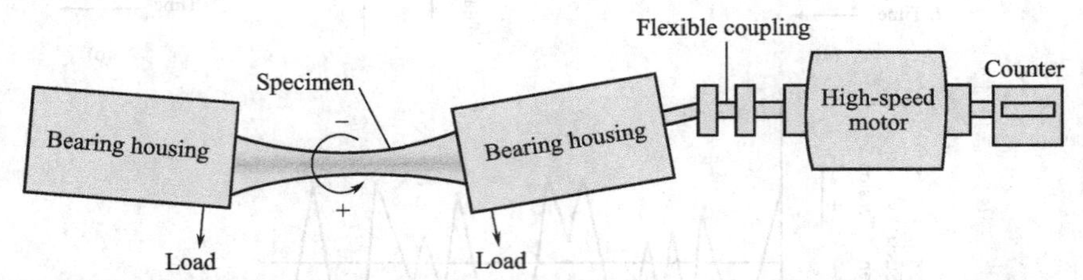

Figure 5.2 Schematic diagram of fatigue-testing apparatus for making rotating bending tests. (From Keyser, Materials Science in Engineering, 4th Edition, © 1986, p. 88. Adapted by permission of Pearson Education, Inc., Upper Saddle River, NJ.)

A series of tests are commenced by subjecting a specimen to the stress cycling at a relatively large maximum stress amplitude (σ_{max}), usually on the order of two thirds of the static tensile strength; the number of cycles to failure is counted. This procedure is repeated on other specimens at progressively decreasing maximum stress amplitudes. Data are plotted as stress S versus the logarithm of the number N of cycles to failure for each of the specimens. The values of S are normally taken as stress amplitudes (σ_a, Equation 5.3); on occasion, σ_{max} or σ_{min} values may be used.

Two distinct types of S–N behavior are observed, which are represented schematically in Figure 5.3. As these plots indicate, the higher the magnitude of the stress, the smaller the number of cycles the material is capable of sustaining before failure. For some ferrous (iron base) and titanium alloys, the S–N curve [Figure 5.3(a)] becomes horizontal at higher N values; or there is a limiting stress level, called the *fatigue limit* (also sometimes the endurance limit), below which fatigue failure will not occur. This fatigue limit represents the largest value of fluctuating stress that will not cause failure for essentially an infinite number of cycles. For many steels, fatigue limits range between 35% and 60% of the tensile strength.

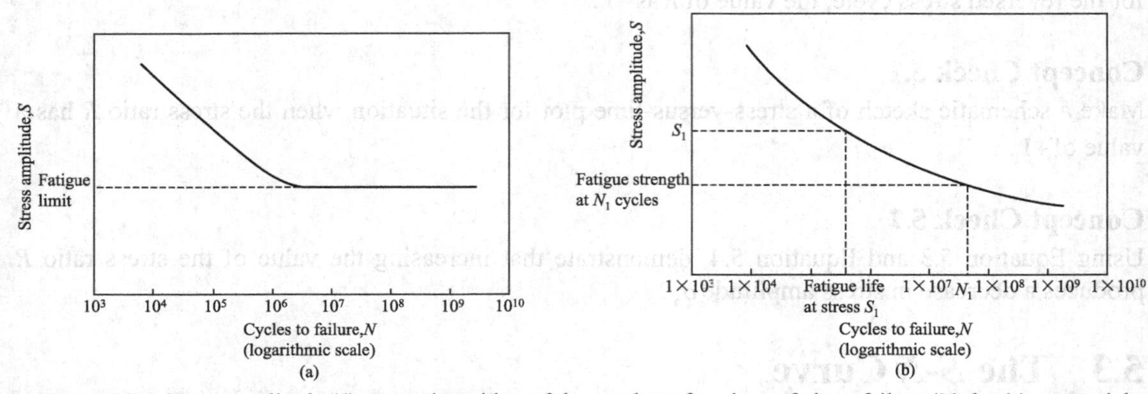

Figure 5.3 Stress amplitude (S) versus logarithm of the number of cycles to fatigue failure (N) for (a) a material that displays a fatigue limit, and (b) a material that does not display a fatigue limit.

Most nonferrous alloys (e.g., aluminum, copper, magnesium) do not have a fatigue limit, in that the *S–N* curve continues its downward trend at increasingly greater *N* values [Figure 5.3(b)].Thus, fatigue will ultimately occur regardless of the magnitude of the stress. For these materials, the fatigue response is specified as fatigue strength, which is defined as the stress level at which failure will occur for some specified number of cycles (e.g., 10^7cycles). The determination of fatigue strength is also demonstrated in Figure 5.3(b).

Another important parameter that characterizes a material's fatigue behavior is fatigue life N_f. It is the number of cycles to cause failure at a specified stress level, as taken from the *S–N* plot [Figure 5.3(b)].

Unfortunately, there always exists considerable scatter in fatigue data—that is, a variation in the measured *N* value for a number of specimens tested at the same stress level. This variation may lead to significant design uncertainties when fatigue life and/or fatigue limit (or strength) is being considered. The scatter in results is a consequence of the fatigue sensitivity to a number of test and material parameters that are impossible to control precisely. These parameters include specimen fabrication and surface preparation, metallurgical variables, specimen alignment in the apparatus, mean stress, and test frequency.

Fatigue *S–N* curves similar to those shown in Figure 5.3 represent "best fit" curves that have been drawn through average-value data points. It is a little unsettling to realize that approximately one-half of the specimens tested actually failed at stress levels lying nearly 25% below the curve (as determined on the basis of statistical treatments).

Several statistical techniques have been developed to specify fatigue life and fatigue limit in terms of probabilities. One convenient way of representing data treated in this manner is with a series of constant probability curves, several of which are plotted in Figure 5.4. The *P* value associated with each curve represents the probability of failure. For example, at a stress of 200 MPa (30000 psi), we would expect 1% of the specimens to fail at about 1×10^6 cycles and 50% to fail at about 2×10^7cycles, and so on. Remember that *S–N* curves represented in the literature are normally average values, unless noted otherwise.

The fatigue behaviors represented in Figure 5.3(a) and Figure 5.3(b) may be classified into two domains. One is associated with relatively high loads that produce not only elastic strain but also some plastic strain during each cycle. Consequently, fatigue lives are relatively short; this domain is

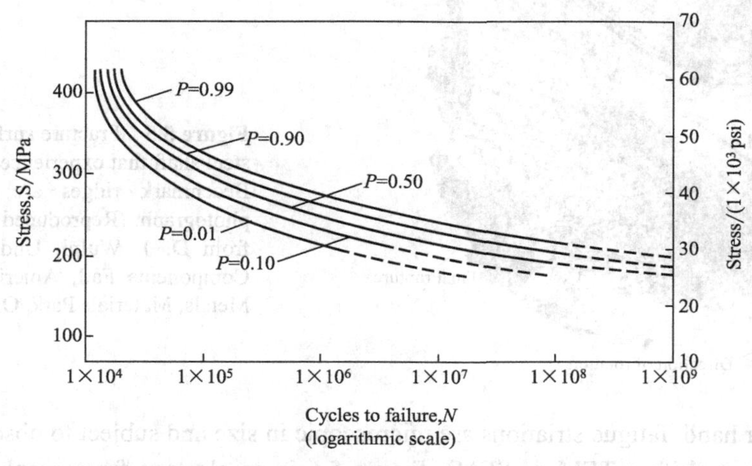

Figure 5.4 Fatigue *S–N* probability of failure curves for a 7075-T6 aluminum alloy; *P* denotes the probability of failure. (From G. M. Sinclair and T. J. Dolan, Trans. ASME, **75,** 1953, p. 867. Reprinted with permission of the American Society of Mechanical Engineers.)

termed low-cycle fatigue and occurs at less than about 1×10^4 to 1×10^5 cycles. For lower stress levels wherein deformations are totally elastic, longer lives result. This is called high-cycle fatigue inasmuch as relatively large numbers of cycles are required to produce fatigue failure. High-cycle fatigue is associated with fatigue lives greater than about 1×10^4 to 1×10^5 cycles.

5.4 Crack Initiation and Propagation

The process of fatigue failure is characterized by three distinct steps:

(1) crack initiation, wherein a small crack forms at some point of high stress concentration;

(2) crack propagation, during which this crack advances incrementally with each stress cycle;

(3) final failure, which occurs very rapidly once the advancing crack has reached a critical size. Cracks associated with fatigue failure almost always initiate (or nucleate) on the surface of a component at some point of stress concentration. Crack nucleation sites include surface scratches, sharp fillets, keyways, threads, dents, and the like. In addition, cyclic loading can produce microscopic surface discontinuities resulting from dislocation slip steps that may also act as stress raisers, and therefore as crack initiation sites.

The region of a fracture surface that formed during the crack propagation step may be characterize8d by two types of markings termed beachmarks and striations. Both of these features indicate the position of the crack tip at some point in time and appear as concentric ridges that expand away from the crack initiation site(s), frequently in a circular or semicircular pattern. Beachmarks (sometimes also called "clamshell marks") are of macroscopic dimensions (Figure 5.5), and may be observed with the unaided eye. These markings are found for components that experienced interruptions during the crack propagation stage—for example, a machine that operated only during normal work-shift hours. Each beachmark band represents a period of time over which crack growth occurred.

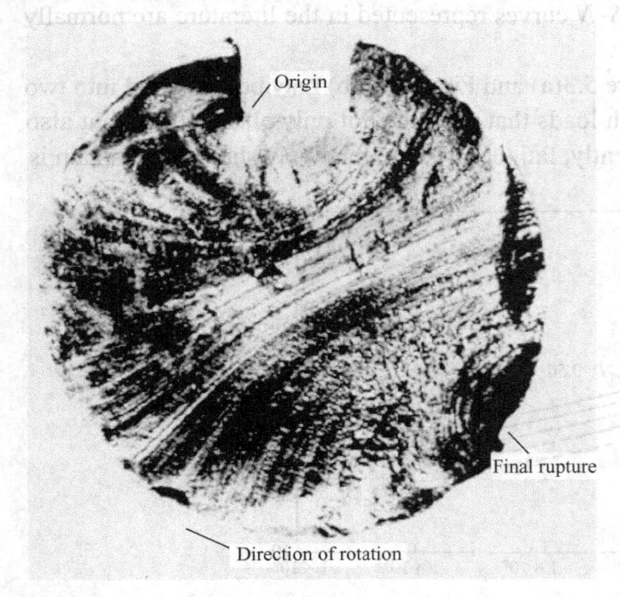

Figure 5.5 Fracture surface of a rotating steel shaft that experienced fatigue failure. Beachmark ridges are visible in the photograph. (Reproduced with permission from D. J. Wulpi, Understanding How Components Fail, American Society for Metals, Materials Park, OH, 1985.)

On the other hand, fatigue striations are microscopic in size and subject to observation with the electron microscope (either TEM or SEM). Figure 5.6 is an electron fractograph that shows this feature. Each striation is thought to represent the advance distance of a crack front during a single load cycle. Striation width depends on, and increases with, increasing stress range.

Figure 5.6 Transmission electron fractograph showing fatigue striations in aluminum. Magnification unknown. (From V. J. Colangelo and F. A. Heiser, Analysis of Metallurgical Failures, 2nd edition. Copyright © 1987 by John Wiley & Sons, New York. Reprinted by permission of John Wiley & Sons, Inc.)

At this point it should be emphasized that although both beachmarks and striations are fatigue fracture surface features having similar appearances, they are nevertheless different, both in origin and size. There may be literally thousands of striations within a single beachmark.

Often the cause of failure may be deduced after examination of the failure surfaces. The presence of beachmarks and/or striations on a fracture surface confirms that the cause of failure was fatigue. Nevertheless, the absence of either or both does not exclude fatigue as the cause of failure.

One final comment regarding fatigue failure surfaces: Beachmarks and striations will not appear on that region over which the rapid failure occurs. Rather, the rapid failure may be either ductile or brittle; evidence of plastic deformation will be present for ductile, and absent for brittle, failure. This region of failure may be noted in Figure 5.7.

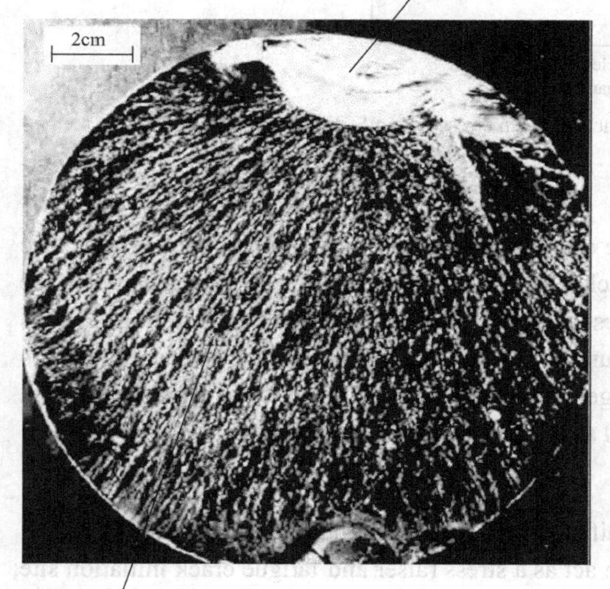

Figure 5.7 Fatigue failure surface. A crack formed at the top edge. The smooth region also near the top corresponds to the area over which the crack propagated slowly. Rapid failure occurred over the area having a dull and fibrous texture (the largest area). Approximately 0.5×. Reproduced by permission from Metals Handbook: Fractography and Atlas of Fractographs, Vol. 9, 8th edition, H. E. Boyer (Editor), American Society for Metals, 1974.]

Concept Check 5.4

Surfaces for some steel specimens that have failed by fatigue have a bright crystalline or grainy appearance. Laymen may explain the failure by saying that the metal crystallized while in service. Offer a criticism for this explanation.

5.5 Factors That Affect Fatigue Life

The fatigue behavior of engineering materials is highly sensitive to a number of variables. Some of these factors include mean stress level, geometrical design, surface effects, and metallurgical variables, as well as the environment. This section is devoted to a discussion of these factors and, in addition, to measures that may be taken to improve the fatigue resistance of structural components.

5.5.1 Mean Stress

The dependence of fatigue life on stress amplitude is represented on the S–N plot. Such data are taken for a constant mean stress σ_m, often for the reversed cycle situation (σ_m=0). Mean stress, however, will also affect fatigue life; this influence may be represented by a series of *S–N* curves, each measured at a different σ_m, as depicted schematically in Figure 5.8. As may be noted, increasing the mean stress level leads to a decrease in fatigue life.

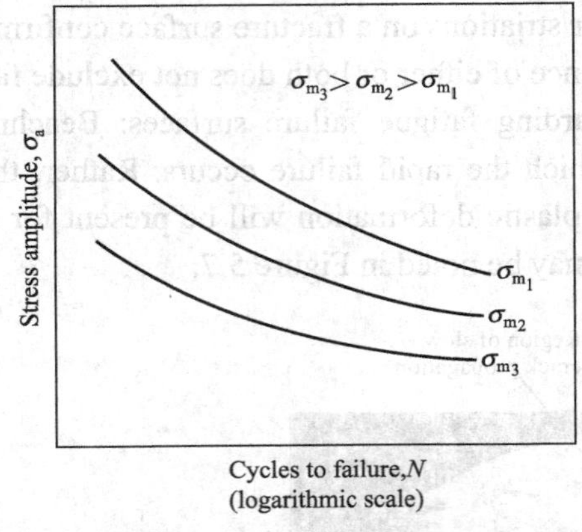

Figure 5.8 Demonstration of the influence of mean stress σ_m on *S–N* fatigue behavior.

5.5.2 Surface Effects

For many common loading situations, the maximum stress within a component or structure occurs at its surface. Consequently, most cracks leading to fatigue failure originate at surface positions, specifically at stress amplification sites. Therefore, it has been observed that fatigue life is especially sensitive to the condition and configuration of the component surface. Numerous factors influence fatigue resistance, the proper management of which will lead to an improvement in fatigue life. These include design criteria as well as various surface treatments.

5.5.3 Design Factors

The design of a component can have a significant influence on its fatigue characteristics.

Any notch or geometrical discontinuity can act as a stress raiser and fatigue crack initiation site; these design features include grooves, holes, keyways, threads, and so on. The sharper the discontinuity (i.e., the smaller the radius of curvature), the more severe the stress concentration. The probability of fatigue failure may be reduced by avoiding (when possible) these structural irregularities, or by making design modifications whereby sudden contour changes leading to sharp corners are eliminated—for example, calling for rounded fillets with large radii of curvature at the point where there is a change in diameter for a rotating shaft (Figure 5.9).

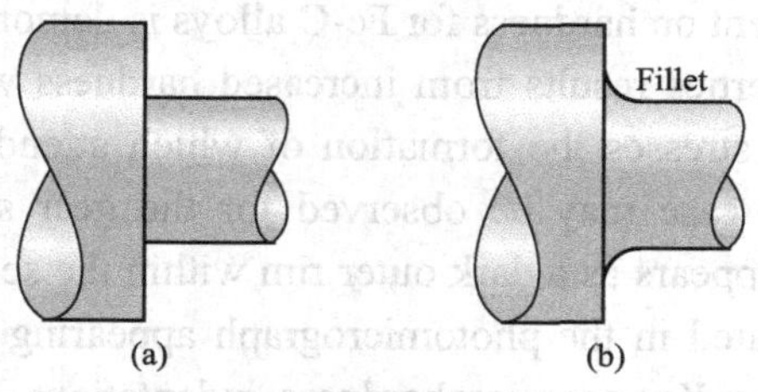

Figure 5.9 Demonstration of how design can reduce stress amplification. (a) Poor design: sharp corner. (b) Good design: fatigue lifetime improved by incorporating rounded fillet into a rotating shaft at the point where there is a change in diameter.

5.5.4 Surface Treatments

During machining operations, small scratches and grooves are invariably introduced into the workpiece surface by cutting tool action. These surface markings can limit the fatigue life. It has been observed that improving the surface finish by polishing will enhance fatigue life significantly.

One of the most effective methods of increasing fatigue performance is by imposing residual compressive stresses within a thin outer surface layer. Thus, a surface tensile stress of external origin will be partially nullified and reduced in magnitude by the residual compressive stress. The net effect is that the likelihood of crack formation and therefore of fatigue failure is reduced.

Residual compressive stresses are commonly introduced into ductile metals mechanically by localized plastic deformation within the outer surface region. Commercially, this is often accomplished by a process termed shot peening. Small, hard particles (shot) having diameters within the range of 0.1mm to 1.0mm are projected at high velocities onto the surface to be treated. The resulting deformation induces compressive stresses to a depth of between one-quarter and one-half of the shot diameter. The influence of shot peening on the fatigue behavior of steel is demonstrated schematically in Figure 5.10.

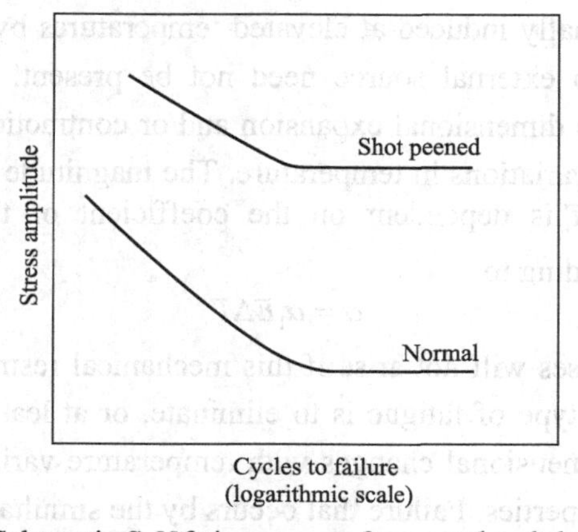

Figure 5.10 Schematic *S–N* fatigue curves for normal and shot-peened steel.

Case hardening is a technique by which both surface hardness and fatigue life are enhanced for steel alloys. This is accomplished by a carburizing or nitriding process whereby a component is exposed to a carbonaceous or nitrogenous atmosphere at an elevated temperature. A carbon- or nitrogen-rich outer surface layer (or "case") is introduced by atomic diffusion from the gaseous phase. The case is normally on the order of 1 mm deep and is harder than the inner core of material (The influence of carbon content on hardness for Fe-C alloys is demonstrated in Figure 10.29a). The improvement of fatigue properties results from increased hardness within the case, as well as the desired residual compressive stresses the formation of which attends the carburizing or nitriding process. A carbon-rich outer case may be observed for the gear shown in the chapter-opening photograph for Chapter 5; it appears as a dark outer rim within the sectioned segment. The increase in case hardness is demonstrated in the photomicrograph appearing in Figure 5.11. The dark and elongated diamond shapes are Knoop microhardness indentations. The upper indentation, lying within the carburized layer, is smaller than the core indentation.

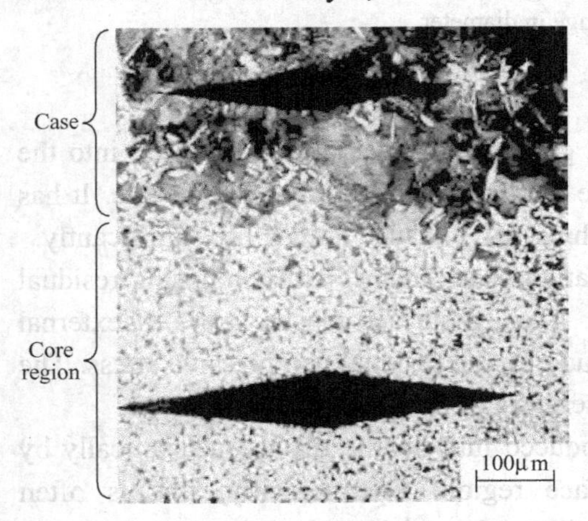

Figure 5.11 Photomicrograph showing both core (bottom) and carburized outer case (top) regions of a casehardened steel. The case is harder as attested by the smaller microhardness indentation. 100×. (From R.W. Hertzberg, Deformation and Fracture)

5.6 Environmental Effects

Environmental factors may also affect the fatigue behavior of materials. A few brief comments will be given relative to two types of environment-assisted fatigue failure: thermal fatigue and corrosion fatigue.

Thermal fatigue is normally induced at elevated temperatures by fluctuating thermal stresses; mechanical stresses from an external source need not be present. The origin of these thermal stresses is the restraint to the dimensional expansion and/or contraction that would normally occur in a structural member with variations in temperature. The magnitude of a thermal stress developed by a temperature change ΔT is dependent on the coefficient of thermal expansion α_l and the modulus of elasticity E according to

$$\sigma = \alpha_l E \Delta T \tag{5.5}$$

Of course, thermal stresses will not arise if this mechanical restraint is absent. Therefore, one obvious way to prevent this type of fatigue is to eliminate, or at least reduce, the restraint source, thus allowing unhindered dimensional changes with temperature variations, or to choose materials with appropriate physical properties. Failure that occurs by the simultaneous action of a cyclic stress and chemical attack is termed corrosion fatigue. Corrosive environments have a deleterious influence and produce shorter fatigue lives. Even the normal ambient atmosphere will affect the fatigue behavior of some materials. Small pits may form as a result of chemical reactions between the environment and material, which serve as points of stress concentration and therefore as crack nucleation sites. In addition, crack propagation rate is enhanced as a result of the corrosive environment. The nature of the stress cycles will influence the fatigue behavior; for example, lowering the load application frequency leads to longer periods during which the opened crack is in contact with the environment and to a reduction in the fatigue life.

Several approaches to corrosion fatigue prevention exist. On one hand, we can take measures to reduce the rate of corrosion by some of the techniques discussed in Chapter 7—for example, apply protective surface coatings, select a more corrosion -resistant material, and reduce the corrosiveness of the environment. And/or it might be advisable to take actions to minimize the probability of normal fatigue failure, as outlined above—for example, reduce the applied tensile stress level and impose residual compressive stresses on the surface of the member.

SUMMARY

Cyclic Stresses (Fatigue)

The *S–N* Curve

Fatigue is a common type of catastrophic failure wherein the applied stress level fluctuates with time. Test data are plotted as stress versus the logarithm of the number of cycles to failure. For many metals and alloys, stress diminishes continuously with increasing number of cycles at failure; fatigue strength and fatigue life are the parameters used to characterize the fatigue behavior of these materials. On the other hand, for other metals/alloys, at some point, stress ceases to decrease with, and becomes independent of, the number of cycles; the fatigue behavior of these materials is expressed in terms of fatigue limit.

Crack Initiation and Propagation

The processes of fatigue crack initiation and propagation were discussed. Cracks normally nucleate on the surface of a component at some point of stress concentration. Two characteristic fatigue surface features are beachmarks and striations. Beachmarks form on components that experience applied stress interruptions; they normally may be observed with the naked eye. Fatigue striations are of microscopic dimensions, and each is thought to represent the crack tip advance distance over a single load cycle.

Factors That Affect Fatigue Life

Measures that may be taken to extend fatigue life include:

(1) reducing the mean stress level;

(2) eliminating sharp surface discontinuities;

(3) improving the surface finish by polishing;

(4) imposing surface residual compressive stresses by shot peening;

(5) case hardening by using a carburizing or nitriding process.

Environmental Effects

The fatigue behavior of materials may also be affected by the environment. Thermal stresses may be induced in components that are exposed to elevated temperature fluctuations and when thermal expansion and/or contraction is restrained; fatigue for these conditions is termed thermal fatigue. The presence of a chemically active environment may lead to a reduction in fatigue life for corrosion fatigue.

IMPORTANT TERMS AND CONCEPTS

Fatigue	Fatigue life	Fatigue limit
Fatigue strength	Thermal fatigue	

REFERENCES

1. Colangelo, V. J. and F. A. Heiser, *Analysis of Metallurgical Failures,* 2nd edition, Wiley, New York, 1987
2. Collins, J.A., *Failure of Materials in Mechanical Design,* 2nd edition,Wiley, New York, 1993
3. Courtney, T. H., *Mechanical Behavior of Materials,* 2nd edition, McGraw-Hill, New York, 2000
4. Dieter, G. E., *Mechanical Metallurgy,* 3rd edition, McGraw-Hill, New York, 1986
5. Esaklul, K. A., *Handbook of Case Histories in Failure Analysis,* ASM International, Materials Park, OH, 1992 and 1993. In two volumes
6. Hertzberg, R. W., *Deformation and Fracture Mechanics of Engineering Materials,* 4th edition, Wiley, New York, 1996
7. Stevens, R. I., A. Fatemi, R. R. Stevens, and H. O. Fuchs, *Metal Fatigue in Engineering,* 2nd edition, Wiley, New York, 2000
8. Tetelman, A. S. and A. J. McEvily, *Fracture of Structural Materials,*Wiley, New York, 1967. Reprinted by Books on Demand, Ann Arbor, MI
9. Wulpi, D. J., *Understanding How Components Fail,* 2nd edition, ASM International, Materials Park, OH, 1999

QUESTIONS AND PROBLEMS

5.1 A fatigue test was conducted in which the mean stress was 70 MPa (10000 psi), and the stress amplitude was 210 MPa (30000 psi).

(a) Compute the maximum and minimum stress levels.

(b) Compute the stress ratio.

(c) Compute the magnitude of the stress range.

5.2 A cylindrical 1045 steel bar (Figure 5.12) is subjected to repeated compression-tension stress cycling along its axis. If the load amplitude is 66700 N (15000 lbf), compute the minimum allowable bar diameter to ensure that fatigue failure will not occur. Assume a safety factor of 2.0.

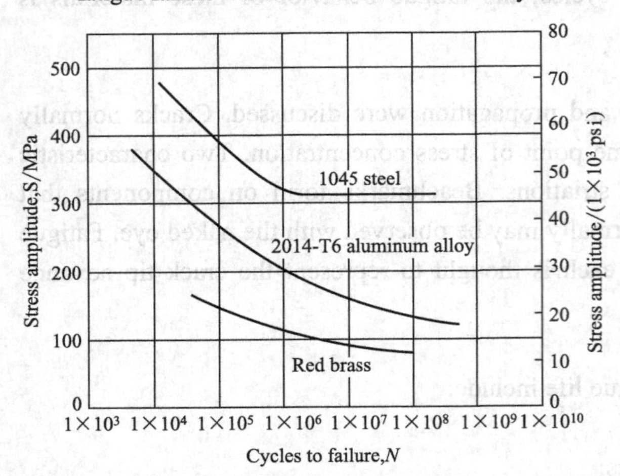

Figure 5.12 Stress magnitude S versus the logarithm of the number N of cycles to fatigue failure for red brass, an aluminum alloy, and a plain carbon steel. (Adapted from H.W. Hayden, W. G. Moffatt, and J.Wulff, The Structure and Properties of Materials, Vol. III, Mechanical Behavior, p. 15.)

5.3 A 6.4 mm (0.25 in) diameter cylindrical rod fabricated from a 2014-T6 aluminum alloy (Figure 8.34) is subjected to reversed tensioncompression load cycling along its axis. If the maximum tensile and compressive loads are +5340 N (+1200 lbf) and −5340 N (−1200 lbf), respectively, determine its fatigue life. Assume that the stress plotted in Figure 8.34 is stress amplitude.

5.4 A 15.2 mm (0.60 in) diameter cylindrical rod fabricated from a 2014-T6 aluminum alloy (Figure 8.34) is subjected to a repeated tension compression load cycling along its axis. Compute the maximum and minimum loads that will be applied to yield a fatigue life of 1.0×10^8 cycles. Assume that the stress plotted on the vertical axis is stress amplitude, and data were taken for a mean stress of 35 MPa (5000 psi).

5.5 The fatigue data for a brass alloy are given as follows:

(a) Make an *S–N* plot (stress amplitude versus logarithm cycles to failure) using these data.
(b) Determine the fatigue strength at 4×10^6 cycles.
(c) Determine the fatigue life for 120 MPa.

Stress Amplitude/MPa	Cycles to Failure	Stress Amplitude/MPa	Cycles to Failure
170	3.7×10^4	92	1.0×10^7
148	1.0×10^5	80	1.0×10^8
130	3.0×10^5	74	1.0×10^9
114	1.0×10^6		

Stress Anmplitude /[MPa(ksi)]	Cycles to Failure	Stress Amplitude /[MPa(ksi)]	Cycles to Failure
470(68.0)	1×10^4	310(45.3)	1×10^6
440(63.4)	3×10^4	290(42.2)	3×10^6
390(56.2)	1×10^5	290(42.2)	1×10^7
350(51.0)	3×10^5	290(42.2)	1×10^8

5.6 Suppose that the fatigue data for the brass alloy in Problem 5.5 were taken from bending-rotating tests, and that a rod of this alloy is to be used for an automobile axle that rotates at an average rotational velocity of 1800 revolutions per minute. Give the maximum torsional stress amplitude possible for each of the following lifetimes of the rod:
(a) 1 year;
(b) 1 month;
(c) 1 day;
(d) 1 hour.

5.7 The fatigue data for a steel alloy are given as follows:
(a) Make an *S–N* plot (stress amplitude versus logarithm cycles to failure) using these data.
(b) What is the fatigue limit for this alloy?
(c) Determine fatigue lifetimes at stress amplitudes of 415 MPa (60000 psi) and 275 MPa (40000 psi).
(d) Estimate fatigue strengths at 2×10^4 and 6×10^5 cycles.

5.8 Suppose that the fatigue data for the steel alloy in Problem 5.7 were taken for bending rotating tests, and that a rod of this alloy is to be used for an automobile axle that rotates at an average rotational velocity of 600 revolutions per minute. Give the maximum lifetimes of continuous driving that are allowable for the following stress levels:
(a) 450 MPa (65000 psi);
(b) 380 MPa (55000 psi);
(c) 310 MPa (45000 psi);
(d) 275 MPa (40000 psi).

5.9 Three identical fatigue specimens (denoted A, B, and C) are fabricated from a nonferrous alloy. Each is subjected to one of the maximum-minimum stress cycles listed below; the frequency is the same for all three tests.

Specimen	σ_{max}/MPa	σ_{min}/MPa
A	+450	–150
B	+300	–300
C	+500	–200

(a) Rank the fatigue lifetimes of these three specimens from the longest to the shortest.
(b) Now justify this ranking using a schematic *S–N* plot.

5.10 Cite five factors that may lead to scatter in fatigue life data.

5.11 Briefly explain the difference between fatigue striations and beach marks both in terms of:

(a) size and;

(b) origin.

5.12 List four measures that may be taken to increase the resistance to fatigue of a metal alloy.

Chapter 5

IMPORTANT TERMS AND CONCEPTS

Fatigue	疲劳	Fatigue Life, N_f	疲劳寿命
Fluctuating Load	变动载荷	Fatigue Strength	疲劳强度
Tension-compression Fatigue	拉压疲劳	Fatigue Limit	疲劳极限
Flexural Fatigue /Bending Fatigue	弯曲疲劳	Crack Initiation	裂纹萌生
Torsional Fatigue /Twisting Fatigue	扭转疲劳	Crack Propagation	裂纹扩展
Cyclic Stress	循环应力	Beachmark	海滩花样
Mean Stress, σ_m	平均应力	Clamshell Mark	贝纹花样
Stress Amplitude, σ_a	应力幅	Striation	条纹状/疲劳线
Stress Ratio, R	应力比	Case Hardening	表面硬化
Low-cycle Fatigue	低周疲劳	Corrosion Fatigue	腐蚀疲劳
High-cycle Fatigue	高周疲劳	Thermal Fatigue	热疲劳
threshold of Fatigue Crack Propagation, ΔK_{th}	疲劳裂纹扩展门槛值		

Chapter 6

Creep of metals

Learning Objectives

After studying this chapter you should be able to do the following:

1. Define creep and specify the conditions under which it occurs.
2. Given a creep plot for some material, determine;

(a) the steady-state creep rate;

(b) the rupture lifetime.

6.1 Introduction

Materials are often placed in service at elevated temperatures and exposed to static mechanical stresses (e.g., turbine rotors in jet engines and steam generators that experience centrifugal stresses, and high-pressure steam lines). Deformation under such circumstances is termed creep. Defined as the time-dependent and permanent deformation of materials when subjected to a constant load or stress, creep is normally an undesirable phenomenon and is often the limiting factor in the lifetime of a part. It is observed in all materials types; for metals it becomes important only for temperatures greater than about $0.4T_m$ (where T_m is absolute melting temperature). Amorphous polymers, which include plastics and rubbers, are especially sensitive to creep deformation as discussed in Section 9.4.

6.2 Generalized Creep Behavior

A typical creep test consists of subjecting a specimen to a constant load or stress while maintaining the temperature constant; deformation or strain is measured and plotted as a function of elapsed time. Most tests are the constant load type, which yield information of an engineering nature; constant stress tests are employed to provide a better understanding of the mechanisms of creep. Figure 6.1 is a schematic representation of the typical constant load creep behavior of metals. Upon application of the load there is an instantaneous deformation, as indicated in the figure, which is mostly elastic. The resulting creep curve consists of three regions, each of which has its own distinctive strain-time feature. Primary or transient creep occurs first, typified by a continuously decreasing creep rate; that is, the slope of the curve diminishes with time. This suggests that the material is experiencing an increase in creep resistance or strain hardening—deformation becomes more difficult as the material is strained. For secondary creep, sometimes termed steady-state creep, the rate is constant; that is, the plot becomes linear. This is often the stage of creep that is of the longest duration. The constancy of creep rate is explained on the basis of a balance between the competing processes of strain hardening and recovery, recovery being the process whereby a material

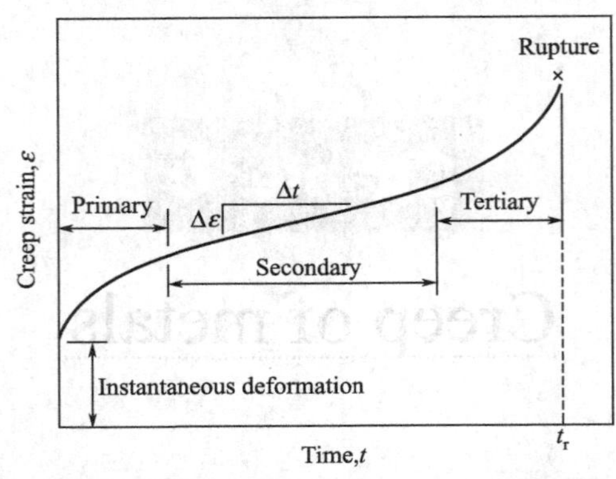

Figure 6.1 Typical creep curve of strain versus time at constant stress and constant elevated temperature. The minimum creep rate $\Delta\varepsilon / \Delta t$ is the slope of the linear segment in the secondary region. Rupture lifetime t_r is the total time to rupture.

becomes softer and retains its ability to experience deformation. Finally, for tertiary creep, there is an acceleration of the rate and ultimate failure. This failure is frequently termed rupture and results from microstructural and/or metallurgical changes; for example, grain boundary separation, and the formation of internal cracks, cavities, and voids. Also, for tensile loads, a neck may form at some point within the deformation region. These all lead to a decrease in the effective cross-sectional area and an increase in strain rate.

For metallic materials most creep tests are conducted in uniaxial tension using a specimen having the same geometry as for tensile tests. On the other hand, uniaxial compression tests are more appropriate for brittle materials; these provide a better measure of the intrinsic creep properties inasmuch as there is no stress amplification and crack propagation, as with tensile loads. Compressive test specimens are usually right cylinders or parallelepipeds having length-to-diameter ratios ranging from about 2 to 4. For most materials creep properties are virtually independent of loading direction.

Possibly the most important parameter from a creep test is the slope of the secondary portion of the creep curve (Figure 6.1); this is often called the minimum or steady-state creep rate It is the engineering design parameter that is considered for long-life applications, such as a nuclear power plant component that is scheduled to operate for several decades, and when failure or too much strain are not options. On the other hand, for many relatively short-life creep situations (e.g., turbine blades in military aircraft and rocket motor nozzles), time to rupture, or the rupture lifetime, is the dominant design consideration; it is also indicated in Figure 6.1. Of course, for its determination, creep tests must be conducted to the point of failure; these are termed creep rupture tests. Thus, a knowledge of these creep characteristics of a material allows the design engineer to ascertain its suitability for a specific application.

Concept Check 6.1

Superimpose on the same strain-versus-time plot schematic creep curves for both constant tensile stress and constant tensile load, and explain the differences in behavior.

6.3 Stress and Temperature Effects

Both temperature and the level of the applied stress influence the creep characteristics (Figure

6.2). At a temperature substantially below $0.4T_m$ and after the initial deformation, the strain is virtually independent of time. With either increasing stress or temperature, the following will be noted:

(1) the instantaneous strain at the time of stress application increases;

(2) the steady-state creep rate is increased;

(3) the rupture lifetime is diminished.

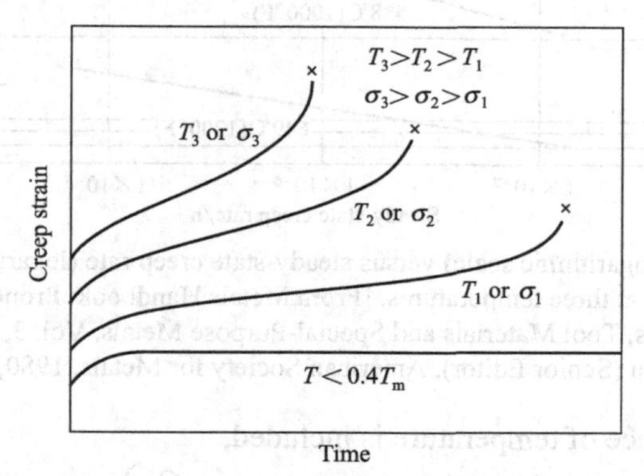

Figure 6.2 Influence of stress σ and temperature T on creep behavior.

The results of creep rupture tests are most commonly presented as the logarithm of stress versus the logarithm of rupture lifetime. Figure 6.3 is one such plot for a nickel alloy in which a linear relationship can be seen to exist at each temperature. For some alloys and over relatively large stress ranges, nonlinearity in these curves is observed.

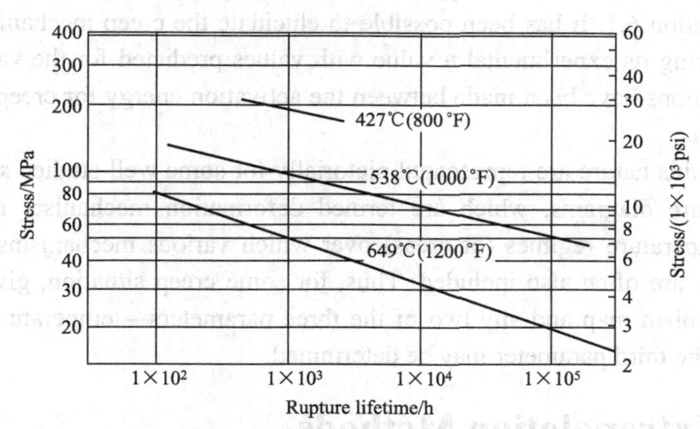

Figure 6.3 Stress (logarithmic scale) versus rupture lifetime (logarithmic scale) for a low carbon–nickel alloy at three temperatures. [From Metals Handbook: Properties and Selection: Stainless Steels, Tool Materials and Special-Purpose Metals, Vol. 3, 9th edition, D. Benjamin (Senior Editor), American Society for Metals, 1980, p. 130.]

Empirical relationships have been developed in which the steady-state creep rate as a function of stress and temperature is expressed. Its dependence on stress can be written

$$\dot{\varepsilon}_s = K_1\sigma^n \tag{6.1}$$

where K_1 and n are material constants. A plot of the logarithm of $\dot{\varepsilon}_s$ versus the logarithm of σ yields a straight line with slope of n; this is shown in Figure 6.4 for a nickel alloy at three temperatures. Clearly, a straight line segment is drawn at each temperature.

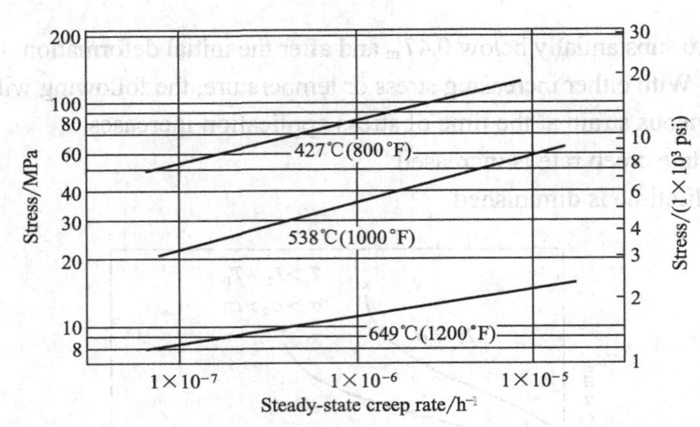

Figure 6.4 Stress (logarithmic scale) versus steady-state creep rate (logarithmic scale) for a low carbon–nickel alloy at three temperatures. [From Metals Handbook: Properties and Selection: Stainless Steels, Tool Materials and Special-Purpose Metals, Vol. 3, 9th edition, D. Benjamin (Senior Editor), American Society for Metals, 1980, p. 131.]

Now, when the influence of temperature is included,

$$\dot{\varepsilon}_s = K_2 \sigma^n \exp\left(-\frac{Q_c}{RT}\right) \tag{6.2}$$

where K_2 and Q_c are constants; Q_c is termed the activation energy for creep.

Several theoretical mechanisms have been proposed to explain the creep behavior for various materials; these mechanisms involve stress-induced vacancy diffusion, grain boundary diffusion, dislocation motion, and grain boundary sliding. Each leads to a different value of the stress exponent n in Equation 6.1. It has been possible to elucidate the creep mechanism for a particular material by comparing its experimental n value with values predicted for the various mechanisms. In addition, correlations have been made between the activation energy for creep and the activation energy for diffusion.

Creep data of this nature are represented pictorially for some well-studied systems in the form of stress–temperature diagrams, which are termed deformation mechanism maps. These maps indicate stress–temperature regimes (or areas) over which various mechanisms operate. Constant strain rate contours are often also included. Thus, for some creep situation, given the appropriate deformation mechanism map and any two of the three parameters—temperature, stress level, and creep strain rate—the third parameter may be determined.

6.4 Data Extrapolation Methods

The need often arises for engineering creep data that are impractical to collect from normal laboratory tests. This is especially true for prolonged exposures (on the order of years). One solution to this problem involves performing creep and/or creep rupture tests at temperatures in excess of those required, for shorter time periods, and at a comparable stress level, and then making a suitable extrapolation to the in-service condition. A commonly used extrapolation procedure employs the Larson–Miller parameter in terms of temperature and rupture lifetime, defined as

$$T(C + \log t_r) \tag{6.3}$$

where C is a constant (usually on the order of 20), for T in Kelvin and the rupture lifetime in hours. The rupture lifetime of a given material measured at some specific stress level will vary with temperature such that this parameter remains constant. Or, the data may be plotted as the logarithm of stress versus the Larson–Miller parameter, as shown in Figure 6.5. Utilization of this technique is demonstrated in the following design example.

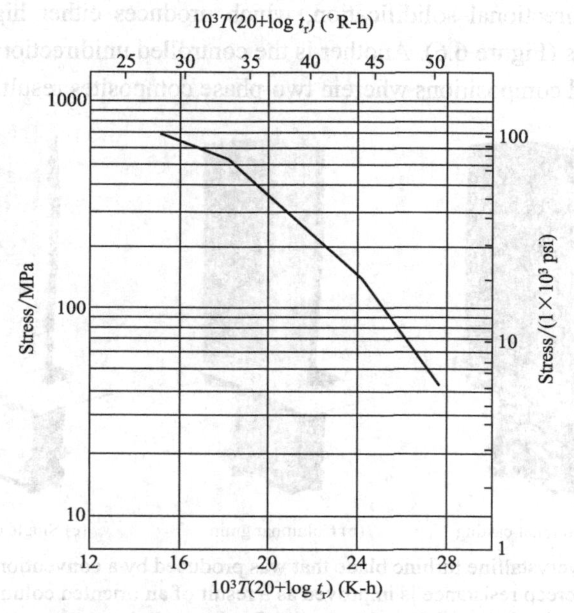

Figure 6.5 Logarithm stress versus the Larson–Miller parameter for an S-590 iron. (From F. R. Larson and J. Miller, *Trans. ASME,* 74, 765, 1952.)

DESIGN EXAMPLE 6.1

Rupture Lifetime Prediction

Using the Larson–Miller data for S-590 iron shown in Figure 6.5, predict the time to rupture for a component that is subjected to a stress of 140 MPa (20000 psi) at 800ºC(1073K).

Solution

From Figure 6.5, at 140 MPa (20000 psi) the value of the Larson–Miller parameter is 24.0×10^3, for T in K and t_r in h; therefore,

$$24.0\times10^3 = T(20+\log t_r) = 1073(20+\log t_r)$$

and, solving for the time

$$22.37 = 20+\log t_r$$

$$t_r = 233\text{h (9.7days)}$$

6.5 Alloys for High-Temperature

There are several factors that affect the creep characteristics of metals. These include melting temperature, elastic modulus, and grain size. In general, the higher the melting temperature, the greater the elastic modulus, and the larger the grain size, the better is a material's resistance to creep. Relative to grain size, smaller grains permit more grain-boundary sliding, which results in higher creep rates. This effect may be contrasted to the influence of grain size on the mechanical behavior at low temperatures [i.e., increase in both strength and toughness].

Stainless steels, the refractory metals, and the superalloys are especially resilient to creep and are commonly employed in high temperature service applications. The creep resistance of the cobalt and nickel superalloys is enhanced by solid-solution alloying, and also by the addition of a dispersed phase that is virtually insoluble in the matrix. In addition, advanced processing techniques have been utilized; one such technique is directional solidification, which produces either highly elongated grains or single-crystal components (Figure 6.6). Another is the controlled unidirectional solidification of alloys having specially designed compositions wherein two-phase composites result.

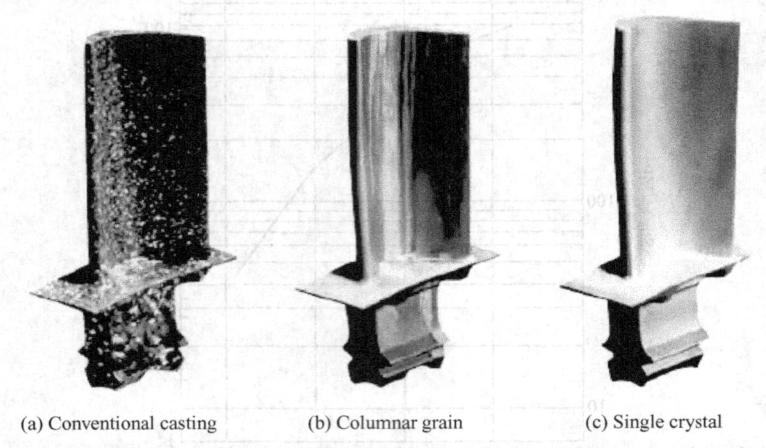

Figure 6.6 (a) Polycrystalline turbine blade that was produced by a conventional casting technique. High-temperature creep resistance is improved as a result of an oriented columnar grain structure (b) produced by a sophisticated directional solidification technique. Creep resistance is further enhanced when single-crystal blades (c) are used.

SUMMARY

Generalized Creep Behavior

The time-dependent plastic deformation of materials subjected to a constant load (or stress) and temperatures greater than about $0.4T_m$ is termed creep. A typical creep curve (strain versus time) will normally exhibit three distinct regions. For transient (or primary) creep, the rate (or slope) diminishes with time. The plot becomes linear (i.e., creep rate is constant) in the steady-state (or secondary) region. And finally, deformation accelerates for tertiary creep, just prior to failure (or rupture). Important design parameters available from such a plot include the steady-state creep rate (slope of the linear region) and rupture lifetime.

Stress and Temperature Effects

Both temperature and applied stress level influence creep behavior. Increasing either of these parameters produces the following effects:

(1) an increase in the instantaneous initial deformation;

(2) an increase in the steady-state creep rate;

(3) a diminishment of the rupture lifetime. Analytical expressions were presented which relate $\dot{\varepsilon}_s$ to both temperature and stress.

Data Extrapolation Methods

Extrapolation of creep test data to lower temperature—longer time regimes is possible using the Larson–Miller parameter.

Alloys for High-Temperature Use

Metal alloys that are especially resistant to creep have high elastic moduli and melting temperatures; these include the superalloys, the stainless steels, and the refractory metals. Various processing techniques are employed to improve the creep properties of these materials.

IMPORTANT TERMS AND CONCEPTS

Creep	Primary or Transient Creep	Tertiary Creep
Rupture	Steady-State Creep Rate	Deformation Mechanism Maps

REFERENCES

1. Colangelo, V. J. and F. A. Heiser, Analysis of Metallurgical Failures, 2nd edition, Wiley, New York, 1987
2. Collins, J.A., Failure of Materials in Mechanical Design, 2nd edition, Wiley, New York, 1993
3. Courtney, T. H., Mechanical Behavior of Materials, 2nd edition, McGraw-Hill, New York, 2000
4. Dieter, G. E., Mechanical Metallurgy, 3rd edition, McGraw-Hill, New York, 1986
5. Esaklul, K. A., Handbook of Case Histories in Failure Analysis, ASM International, Materials Park, OH, 1992 and 1993. In two volumes
6. Hertzberg, R. W., Deformation and Fracture Mechanics of Engineering Materials, 4th edition, Wiley, New York, 1996
7. Stevens, R. I., A. Fatemi, R. R. Stevens, and H. O. Fuchs, Metal Fatigue in Engineering, 2nd edition, Wiley, New York, 2000
8. Tetelman, A. S. and A. J. McEvily, Fracture of Structural Materials, Wiley, New York, 1967. Reprinted by Books on Demand, Ann Arbor, MI
9. Wulpi, D. J., Understanding How Components Fail, 2nd edition, ASM International, Materials Park, OH, 1999

QUESTIONS AND PROBLEMS

6.1 Give the approximate temperature at which creep deformation becomes an important consideration for each of the following metals: tin, molybdenum, iron, gold, zinc, and chromium.

6.2 The following creep data were taken on an aluminum alloy at 480°C (900°F) and a constant stress of 2.75 MPa (400 psi). Plot the data as strain versus time, then determine the steady-state or minimum creep rate. Note: The initial and instantaneous strain is not included.

Time/min	Strain	Time/min	Strain
0	0.00	18	0.82
2	0.22	20	0.88
4	0.34	22	0.95
6	0.41	24	1.03
8	0.48	26	1.12
10	0.55	28	1.22
12	0.62	30	1.36
14	0.68	32	1.53
16	0.75	34	1.77

6.3 A specimen 1015 mm (40 in) long of a low carbon–nickel alloy (Figure 6.6) is to be exposed to a tensile stress of 70 MPa (10000 psi) at 427°C (800°F). Determine its elongation after 10000 h. Assume that the total of both instantaneous and primary creep elongations is 1.3 mm (0.05 in).

6.4 For a cylindrical low carbon–nickel alloy specimen (Figure 6.6) originally 19 mm (0.75 in) in diameter and 635 mm (25 in) long, what tensile load is necessary to produce a total elongation of 6.44 mm (0.25 in) after 5000 h at 538°C (1000°F). Assume that the sum of instantaneous and primary creep elongations is 1.8 mm (0.07 in).

6.5 If a component fabricated from a low carbon–nickel alloy (Figure 6.5) is to be exposed to a tensile stress of 31 MPa (4500 psi) at 649 °C (1200°F), estimate its rupture lifetime.

6.6 A cylindrical component constructed from a low carbon–nickel alloy (Figure 6.5) has a diameter of 19.1 mm

(0.75 in). Determine the maximum load that may be applied for it to survive 10000 h at 538 ℃ (1000℉).

6.7 From Equation 8.19, if the logarithm of $\dot{\varepsilon}_s$ is plotted versus the logarithm of σ, then a straight line should result, the slope of which is the stress exponent n. Using Figure 6.6, determine the value of n for the low carbon–nickel alloy at each of the three temperatures.

6.8 (a) Estimate the activation energy for creep(i.e., Q_c in Equation 8.20) for the low carbon –nickel alloy having the steady-state creep behavior shown in Figure 6.6. Use data taken at a stress level of 55 MPa (8000 psi) and temperatures of 427 ℃ and 538 ℃. Assume that the stress exponent n is independent of temperature.

(b) Estimate $\dot{\varepsilon}_s$ at 649℃ (922 K).

6.9 Steady-state creep rate data are given here for some alloy taken at 200 ℃ (473 K):

$\dot{\varepsilon}_s$ /h^{-1}	σ/[MPa(psi)]
2.5×10^{-3}	55(8000)
2.4×10^{-2}	69(10000)

If it is known that the activation energy for creep is 140000 J/mol, compute the steady-state creep rate at a temperature of 250℃ (523 K) and a stress level of 48 MPa (7000 psi).

6.10 Steady-state creep data taken for an iron at a stress level of 140 MPa (20000 psi) are given here:

$\dot{\varepsilon}_s$ /h^{-1}	T/K
6.6×10^{-4}	1090
8.8×10^{-2}	1200

If it is known that the value of the stress exponent n for this alloy is 8.5, compute the steady-state creep rate at 1300 K and a stress level of 83 MPa (12000 psi).

6.11 Cite three metallurgical/processing techniques that are employed to enhance the creep resistance of metal alloys.

Chapter 6

IMPORTANT TERMS AND CONCEPTS

Creep	蠕变	Primary Creep	第一阶段蠕变
Transient Creep	过渡蠕变	Secondary Creep	第二阶段蠕变
Steady-State Creep	稳态蠕变	Tertiary Creep	第三阶段蠕变
Steady-State Creep Rate	稳态蠕变速率	Instantaneous deformation	瞬时变形
Rupture	断裂	Rupture lifetime	破裂寿命
Decreasing-rate creep	减速蠕变	Creep strength	蠕变强度
Constant-rate creep	恒速蠕变	Creep limit	蠕变极限
Acceleration-rate creep	加速蠕变	creep rupture strength /endurance strength	持久强度
Strain rate	应变速率		

Chapter 7

Corrosion and Degradation of Metals

Learning Objectives

After careful study of this chapter you should be able to do the following:

1. Understanding the electrochemical nature of corrosion in Metals.
2. Briefly explain passivity phenomenon.
3. Define stress corrosion cracking and explain its mechanic.
4. Explain the mechanic of hydrogen damage in Metals.

7.1 Introduction

Environment by its omnipresence, except perhaps in space, affects the behavior of all materials. Such effects can range from swelling in polymers to surface oxidation of metals and nonoxide ceramics to catastrophic failure of some materials under a combined action of stress and environment. Environmental degradation of materials is often referred to as corrosion. Such damage is generally time dependent, and, one is able to predict it. Over time, however, environmental damage can become critical. There is, however, a more insidious corrosion problem which is time-independent. Examples of time-independent corrosion include stress corrosion cracking (SCC), environment induced embrittlement, etc. Such damage can occur at anytime, without much warning. There are many examples of such failures resulting in human and economic loss. Corrosion of structural components in aging aircraft is a serious problem. Just to cite one such example, a Boeing 737 belonging to Aloha Airlines, which flies inter island in Hawaii, lost a large portion of its upper fuselage at 7500 m (24000 feet) in the air. It turned out that the fuselage panels joined by rivets had corroded, which resulted in the mid-flight failure due to corrosion fatigue.

All materials (metals, ceramics, and polymers) show phenomena of premature failure or mechanical property degradation under certain combinations of stress and environment. We describe below the salient points in regard to environmental effects in different materials. We emphasize the role that the microstructure of a given material plays in this phenomenon, especially in environmentally assisted fracture.

7.2 Electrochemical Nature of Corrosion in Metals

For metallic materials, the corrosion process is normally electrochemical, that is, a chemical reaction in which there is transfer of electrons from one chemical species to another. Metal atoms characteristically lose or give up electrons in what is called an oxidation reaction. For example, the hypothetical metal M that has a valence of n (or n valence electrons) may experience oxidation according to the reaction

$$M \rightarrow M^{n+} + ne^- \quad (7.1)$$

in which M becomes an $n+$ positively charged ion and in the process loses its n valence electrons; e^- is used to symbolize an electron. Examples in which metals oxidize are

$$Fe \rightarrow Fe^{2+} + 2e^- \quad (7.2a)$$

$$Al \rightarrow Al^{3+} + 3e^- \quad (7.2b)$$

The site at which oxidation takes place is called the anode; oxidation is sometimes called an anodic reaction.

The electrons generated from each metal atom that is oxidized must be transferred to and become a part of another chemical species in what is termed a reduction reaction. For example, some metals undergo corrosion in acid solutions, which have a high concentration of hydrogen (H^+) ions; the H^- ions are reduced as follows:

$$2H^+ + 2e^- \rightarrow H_2 \quad (7.3)$$

and hydrogen gas (H_2) is evolved.

Other reduction reactions are possible, depending on the nature of the solution to which the metal is exposed. For an acid solution having dissolved oxygen, reduction according to

$$O_2 + 4H^+ + 4e^- \rightarrow 2H_2O \quad (7.4)$$

will probably occur. Or, for a neutral or basic aqueous solution in which oxygen is also dissolved,

$$O_2 + 4H^+O + 4e^- \rightarrow 4(OH^-) \quad (7.5)$$

Any metal ions present in the solution may also be reduced; for ions that can exist in more than one valence state (multivalent ions), reduction may occur by

$$M^{n+} + e^- \rightarrow M^{(n-1)+} \quad (7.6)$$

in which the metal ion decreases its valence state by accepting an electron. Or a metal may be totally reduced from an ionic to a neutral metallic state according to

$$M^{n+} + ne^- \rightarrow M \quad (7.7)$$

The location at which reduction occurs is called the cathode. Furthermore, it is possible for two or more of the reduction reactions above to occur simultaneously. An overall electrochemical reaction must consist of at least one oxidation and one reduction reaction, and will be the sum of them; often the individual oxidation and reduction reactions are termed half-reactions. There can be no net electrical charge accumulation from the electrons and ions; that is, the total rate of oxidation must equal the total rate of reduction, or all electrons generated through oxidation must be consumed by reduction.

For example, consider zinc metal immersed in an acid solution containing H^+ ions. At some regions on the metal surface, zinc will experience oxidation or corrosion as illustrated in Figure 7.1, and according to the reaction

$$Zn \rightarrow Zn^{2+} + 2e^- \quad (7.8)$$

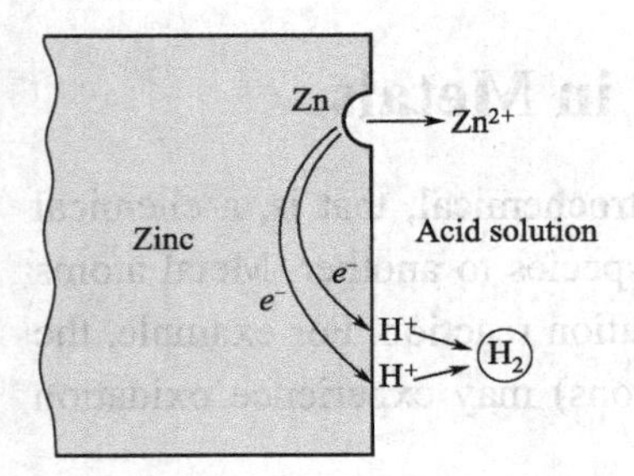

Figure 7.1 The electrochemical reactions associated with the corrosion of zinc in an acid solution. (From M. G Fontana, Corrosion Engineering, 3rd edition.)

Since zinc is a metal, and therefore a good electrical conductor, these electrons may be transferred to an adjacent region at which the H^+ ions are reduced according to

$$2H^+ + 2e^- \rightarrow H_2(\uparrow) \tag{7.9}$$

If no other oxidation or reduction reactions occur, the total electrochemical reaction is just the sum of reaction 7.8 and reaction 7.9, or

$$Zn + 2H^+ \rightarrow Zn^{2+} + H_2(\uparrow) \tag{7.10}$$

Another example is the oxidation or rusting of iron in water, which contains dissolved oxygen. This process occurs in two steps; in the first, Fe is oxidized to Fe^{2+}[as $Fe(OH)_2$],

$$Fe + \frac{1}{2}O_2 + H_2O \rightarrow Fe^{2+} + 2OH^- \rightarrow Fe(OH)_2 \tag{7.11}$$

and, in the second stage, to Fe^{3+} [as $Fe(OH)_3$] according to

$$2Fe(OH)_2 + \frac{1}{2}O_2 + H_2O \rightarrow 2Fe(OH)_3 \tag{7.12}$$

The compound $Fe(OH)_3$ is the all too familiar rust.

As a consequence of oxidation, the metal ions may either go into the corroding solution as ions (reaction 7.8), or they may form an insoluble compound with nonmetallic elements as in Figure 7.2.

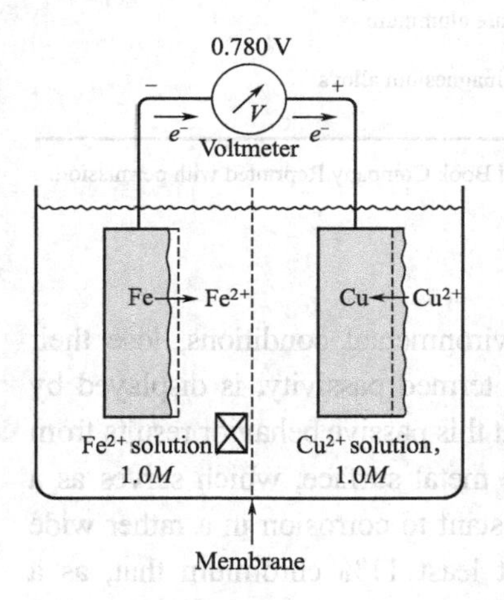

Figure 7.2 An electrochemical cell consisting of iron and copper electrodes, each of which is immersed in a 1*M* solution of its ion. Iron corrodes while copper electrodeposits.

Even though Table 7.1 was generated under highly idealized conditions and has limited utility, it nevertheless indicates the relative reactivities of the metals. A more realistic and practical ranking, however, is provided by the galvanic series, Table 7.1. This represents the relative reactivities of a number of metals and commercial alloys in seawater. The alloys near the top are cathodic and unreactive, whereas those at the bottom are most anodic; no voltages are provided.

Most metals and alloys are subject to oxidation or corrosion to one degree or another in a wide variety of environments; that is, they are more stable in an ionic state than as metals. In thermodynamic terms, there is a net decrease in free energy in going from metallic to oxidized states. Consequently, essentially all metals occur in nature as compounds—for example, oxides, hydroxides, carbonates, silicates, sulfides, and sulfates. Two notable exceptions are the noble metals gold and platinum.For them, oxidation in most environments is not favorable, and, therefore, they may exist in nature in the metallic state.

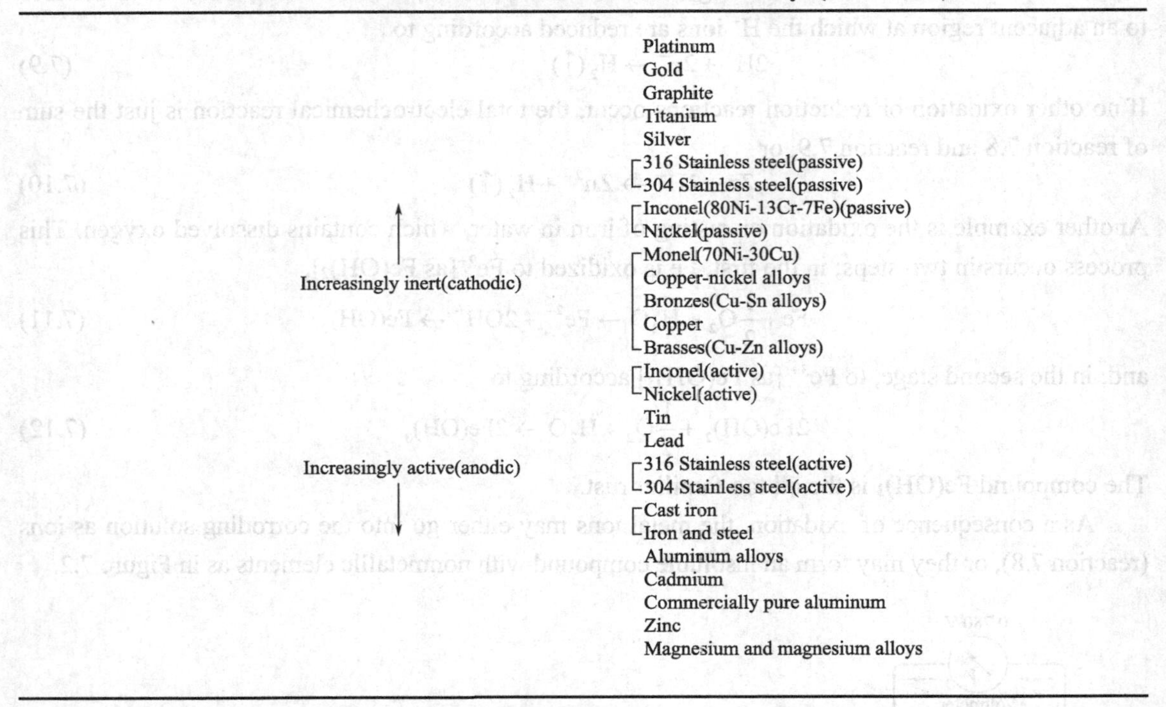

Table 7.1 The Galvanic Series of some metals and alloys (in seawater)

Trend	Metal or alloy
	Platinum
	Gold
	Graphite
	Titanium
	Silver
	316 Stainless steel(passive)
	304 Stainless steel(passive)
	Inconel(80Ni-13Cr-7Fe)(passive)
↑	Nickel(passive)
	Monel(70Ni-30Cu)
Increasingly inert(cathodic)	Copper-nickel alloys
	Bronzes(Cu-Sn alloys)
	Copper
	Brasses(Cu-Zn alloys)
	Inconel(active)
	Nickel(active)
	Tin
	Lead
Increasingly active(anodic)	316 Stainless steel(active)
	304 Stainless steel(active)
↓	Cast iron
	Iron and steel
	Aluminum alloys
	Cadmium
	Commercially pure aluminum
	Zinc
	Magnesium and magnesium alloys

Source:M.G.Fontana,Corrosion Engineering,3rd edition.Copyright 1986 by McGraw-Hill Book Company.Reprinted with permission.

7.3 Passivity

Some normally active metals and alloys, under particular environmental conditions, lose their chemical reactivity and become extremely inert. This phenomenon, termed passivity, is displayed by chromium, iron, nickel, titanium, and many of their alloys. It is felt that this passive behavior results from the formation of a highly adherent and very thin oxide film on the metal surface, which serves as a protective barrier to further corrosion. Stainless steels are highly resistant to corrosion in a rather wide variety of atmospheres as a result of passivation. They contain at least 11% chromium that, as a solid-solution alloying element in iron, minimizes the formation of rust; instead, a protective surface film forms in oxidizing atmospheres. (Stainless steels are susceptible to corrosion in some environments, and therefore are not always "stainless".) Aluminum is highly corrosion resistant in many environments because it also passivates. If damaged, the protective film normally reforms very rapidly. However, a change in the character of the environment (e.g., alteration in the concentration of the active corrosive species) may cause a passivated material to revert to an active state. Subsequent damage to a preexisting passive film could result in a substantial increase in corrosion rate, by as much as 100000 times.

This passivation phenomenon may be explained in terms of polarization potential–log current density curves discussed in the preceding section. The polarization curve for a metal that passivates will have the general shape shown in Figure 7.3. At relatively low potential values, within the "active" region the behavior is linear as it is for normal metals. With increasing potential, the current density suddenly decreases to a very low value that remains independent of potential; this is termed the "passive" region. Finally, at even higher potential values, the current density again increases with potential in the "transpassive" region.

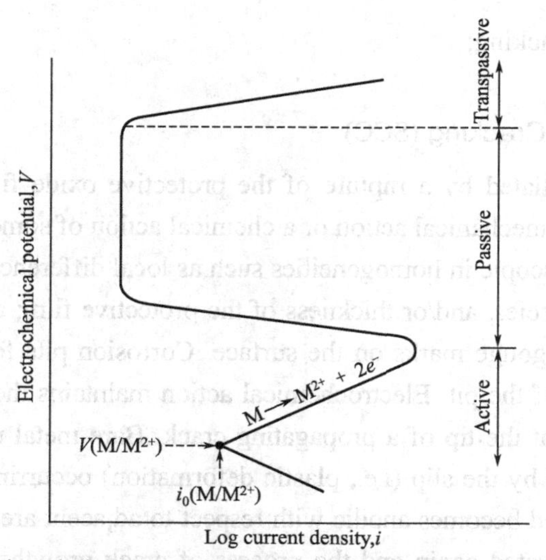

Figure 7.3 Schematic polarization curve for a metal that displays an active–passive transition.

Figure 7.4 illustrates how a metal can experience both active and passive behavior depending on the corrosion environment. Included in this figure is the S-shaped oxidation polarization curve for an active–passive metal M and, in addition, reduction polarization curves for two different solutions, which are labeled 1 and 2. Curve 1 intersects the oxidation polarization curve in the active region at point *A*, yielding a corrosion current density $i_c(A)$. The intersection of curve 2 at point *B* is in the passive region and at current density $i_c(B)$. The corrosion rate of metal M in solution 1 is greater than in solution 2 since $i_c(A)$ is greater than $i_c(B)$ and rate is proportional to current density. This difference in corrosion rate between the two solutions may be significant —several orders of magnitude—when one consider that the current density scale in Figure 7.4 is scaled logarithmically.

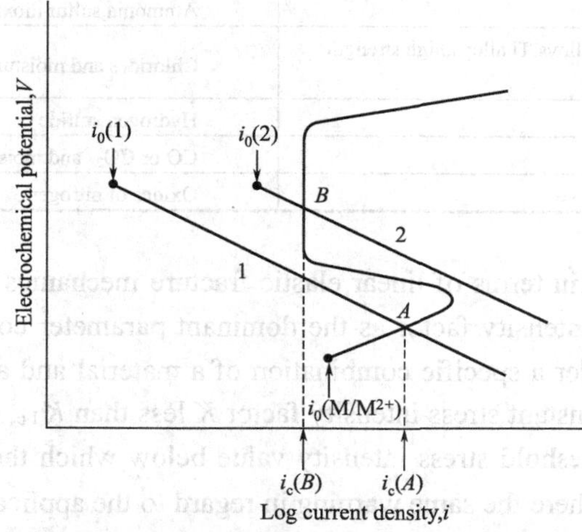

Figure 7.4 Demonstration of how an active–passive metal can exhibit both active and passive corrosion behaviors.

7.4 Environmentally Assisted Fracture in Metals

Environmentally assisted fracture in metals can be classified under the following subheadings:

(1) stress corrosion cracking;

(2) hydrogen damage.

7.4.1 Stress Corrosion Cracking (SCC)

Generally, SCC is initiated by a rupture of the protective oxide film on the metal. This film rupture may occur due to a mechanical action or a chemical action of some species. Possible initiation sites of SCC include microscopic in homogeneities such as local differences in chemical composition, amount of the corrosive species, and/or thickness of the protective film; and any stress concentration sites such as a preexisting gouge marks on the surface. Corrosion pits form at the rupture sites and cracking starts at the root of the pit. Electrochemical action maintains the sharpness of the crack tip, with corrosion continuing at the tip of a propagating crack. Bare metal under the protective film or passivated layer is exposed by the slip (i.e., plastic deformation) occurring at the crack tip. The new metal surface that is exposed becomes anodic with respect to adjacent areas that act cathodically. The corroding metal gets passivated again and the process of crack growth is repeated. The crack thus propagates in a stepwise manner in a transgranular or intergranular mode depending on the metal and environmental conditions. Characteristically, SCC shows branching, with the main crack growing in a direction perpendicular to the major tensile stress component and a low ductility.

As mentioned above, SCC occurs under the combined action of a tensile stress (applied or residual) and an aggressive environment. However, a specific metal/environment combination is required for SCC to occur. Examples include aluminum alloys/seawater, brass/ammonia, austenitic stainless steel/seawater, titanium/liquid nitrogen tetroxide (N_2O_4), etc. Table 7.2 summarizes some of the important metal/environment combinations.

Table 7.2 Some Important Alloy/Environment Combinations for SCC

Alloys	Environments
Copper alloys	Ammonia,sulfur dioxide,oxygen
Austenitic stainless steels,Al alloys,Ti alloys,high strength steels	Chlorides and moisture
Low carbon steels	Hydrogen sulfide
Carbon steels	CO or CO_2 and moisture
Copper alloys	Oxides of nitrogen

The treatment of SCC in terms of linear elastic fracture mechanics (LEFM) analysis involves the use of crack-tip stress intensity factor as the dominant parameter controlling the crack growth under SCC conditions. Under a specific combination of a material and an aggressive environment, cracks can grow under a constant stress intensity factor K less than $K_{\mathrm{I c}}$, the fracture toughness. We then define $K_{\mathrm{I\,scc}}$ as the threshold stress intensity value below which the crack propagation rate is negligible. One should add here the same warning in regard to the applicability of the linear fracture mechanics concepts as was done in the case of ordinary fracture in the absence of an aggressive environment; that is the size of the plastic zone at the crack tip must be small compared to the specimen dimensions for the application of LEFM to be valid. Crack growth velocity varies with the stress intensity factor, K. A schematic plot of log da/dt vs. applied stress intensity is shown in Figure 7.5. There are three regions in this curve:

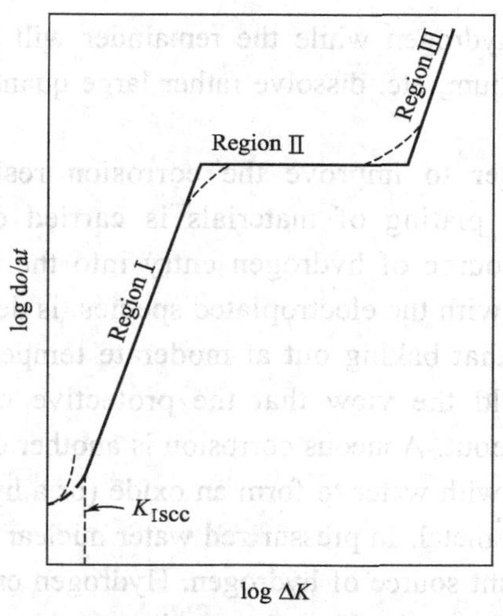

Figure 7.5 Crack growth rate as function of the stress intensity factor under conditions of SCC.

Region Ⅰ: In this region the crack velocity depends on the stress intensity factor. The threshold stress intensity, $K_{I\,scc}$, below which the crack growth does not occur, is shown by a dashed line. Quite frequently, a true $K_{I\,scc}$ does not exist. In such a case, we can define an operational $K_{I\,scc}$ as that corresponding to a crack growth rate of 1×10^{-9} or 1×10^{-10} ms^{-1}. Such an arbitrary value can be used to rate different alloys.

Region Ⅱ: The crack velocity in this region is independent of the stress intensity factor. The value at which this plateau region occurs is very specific to metal/environment combination and test conditions such as temperature.

Region Ⅲ: In this region the crack velocity becomes very fast as the cracktip stress intensity factor approaches K_{Ic}. In this region, the crack velocity is mainly controlled by the stress intensity.

7.4.2 Hydrogen Damage in Metals

The presence of hydrogen in a material can cause serious damage to its performance. In addition to its great technological importance, the phenomenon of hydrogen damage has been a challenging basic research problem. One main reason for the damage caused by hydrogen in metals and alloys is the extremely small size of the hydrogen atom, which makes it move very fast in the metallic lattice. It is therefore not surprising that over the years a considerable research effort has gone into obtaining an understanding of the phenomenon, especially in metals and alloys. We provide below a short account of the hydrogen effects in various metals and alloys.

Some of the common sources of hydrogen in metals as well as some simple and straightforward remedies for the problem are as follows. Metals may absorb hydrogen during processing or service. For example, during melting and casting of metals, the hot metal can react with the raw materials or the humidity in air to form an oxide and hydrogen. The latter can be absorbed by the hot metal. This problem of hydrogen absorption by the liquid metal can be reduced by vacuum degassing processing.

Atmospheric humidity can be a source of hydrogen in the arc welding of steels, while the electrode itself may absorb hydrogen during casting. Frequently, during some steps in the processing of a metal into a useful article, a chemical or electrochemical treatment is given. Nascent-hydrogen is released due to reaction of metal with acid during such a treatment. Most of it combines to form molecular hydrogen while the remainder will diffuse into the metal. Certain metals such as titanium, zirconium, etc. dissolve rather large quantities of hydrogen exothermally and form very brittle hydrides.

Quite frequently, in order to improve the corrosion resistance and/or for decorative purposes, electropolishing or plating of materials is carried out. Such finishing processes represent another important source of hydrogen entry into the base metal. In these finishing processes, hydrogen, together with the electroplated species, is deposited at the cathode. In such cases, it is thought by some that baking out at moderate temperatures after plating may help remove hydrogen. Others hold the view that the protective coating serves as a barrier to hydrogen removal during bakeout. Aqueous corrosion is another common source of hydrogen for metals in service. Metal reacts with water to form an oxide (or a hydroxide) and atomic hydrogen, which is easily absorbed in the metal. In pressurized water nuclear reactors (PWR), water used for heat transfer can be an important source of hydrogen. Hydrogen embrittlement of zirconium alloy fuel cladding or of the pressure vessel itself can be a serious problem.

In the chemical and petrochemical industry, containers of chemicals (used for storage or as reaction chambers) can absorb hydrogen over a period of use. Natural gas containing H_2S, called sour gas, can cause hydrogen induced cracking (HIC) in the pipeline steel. The sulfide ion is especially a problem species because it acts as a ‘‘surface poison” retarding the recombination of atomic hydrogen to form molecular hydrogen at the surface, leading to absorption of atomic hydrogen.

Theories of Hydrogen Damage

No single model or theory is capable of explaining all the effects associated with the presence of hydrogen in different materials. However, almost all theories recognize that one of the most important attributes of hydrogen is that it diffuses very rapidly in most any material. For example, in steels hydrogen diffuses about 10 μm per second at room temperature. This fast diffusion characteristic of hydrogen stems partly from its extremely small size; for hydrogen has the smallest atomic diameter among all the elements. In general, hydrogen tends to collect at defect sites in any material where it can produce high internal pressure, which can lead to cracking. There are certain special aspects of the hydrogen behavior in steels. Hydrogen has a very high mobility in the BCC lattice of Fe at ambient temperature. The comparative values of the diffusivity of hydrogen and nitrogen in the iron lattice at room temperature given below give a good idea of the extraordinarily high mobility of the hydrogen atom.

$$D_H \text{ in Fe} \sim 10^{-2}\ \mathrm{m^2\ s^{-1}} \text{ at } 300\ \mathrm{K}$$

$$D_N \text{ in Fe} \sim 10^{-12}\ \mathrm{m^2\ s^{-1}} \text{ at } 300\ \mathrm{K}$$

One can write for the local concentration of hydrogen in the BCC iron lattice as:

$$\ln \frac{C_H}{C_0} = \frac{\Omega \sigma_p}{RT}$$

where C_H is the local hydrogen concentration, C_0 is the equilibrium hydrogen concentration in the unstressed lattice, Ω is the molar volume of hydrogen in iron, σ_p is the hydrostatic stress [$= (\sigma_1+\sigma_2+\sigma_3)/3$]. Thus, in any nonuniformly stressed solid, there is a driving force for solute migration, which is a function of the solute atomic volume and the gradient in the hydrostatic stress component of the applied stress. Hydrogen segregates to regions of large hydrostatic tension. Figure 7.6 shows schematically the transport processes at a cracktip that eventually lead to the embrittlement reaction between the hydrogen and the metal, in this case iron.

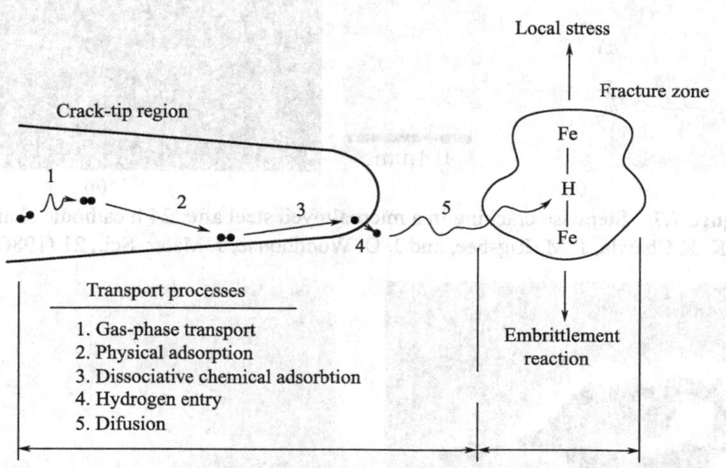

Figure 7.6 Schematic of the hydrogen transport processes at a crack tip in Fe and the embrittlement reaction [After R. P. Gangloff and R. P. Wei, Met. Trans. A, 8A (1977)1043].

This hydrogen transport process can be divided into the following steps:

(a) diffusion of hydrogen to the surface;

(b) adsorption on the surface;

(c) dissociation in the surface adsorption layer;

(d) penetration through the surface;

(e) diffusion into the bulk of the metal.

Having given this very general picture of the effects of hydrogen in metals, we review briefly some of the specific theories that have been advanced to explain the phenomenon of hydrogen damage.

Lattice Decohesion

A hydrogen induced lattice decohesion can occur as originally proposed by Toriano. Hydrogen diffuses into the triaxial tensile stress region at a crack tip, causing a localized reduction of the lattice cohesive strength. The concept is quite valid in very general terms. The exact mechanisms involved are, however, not clear.

Pressure theory

Hydrogen atoms combine and precipitate as molecular hydrogen and cause internal pressure. When this internal pressure exceeds a critical value, HIC occurs. Because of the extremely high mobility of hydrogen in most lattices, segregation of absorbed hydrogen to regions of high expansion in the lattice, for example, internal voids and cracks, occurs easily. Large internal pressure would enhance void growth and crack propagation. A good example of this phenomenon is the blister formation in steels on cathodic charging. One would expect such cracking to vary with inclusion distribution. Figure 7.7 shows such hydrogen induced cracking in a microalloyed steel sample. This extensive stepwise cracking resulted after cathodic charging for 24 h. Such cracking or voiding is frequently associated with the presence of inclusions.

Figure 7.8 shows an aluminum-based inclusion (possibly alumina) in the interior of a void produced by hydrogen charging. The micrograph on the right in Figure 7.8 shows the mapping of aluminum, indicating an aluminum-based inclusion.

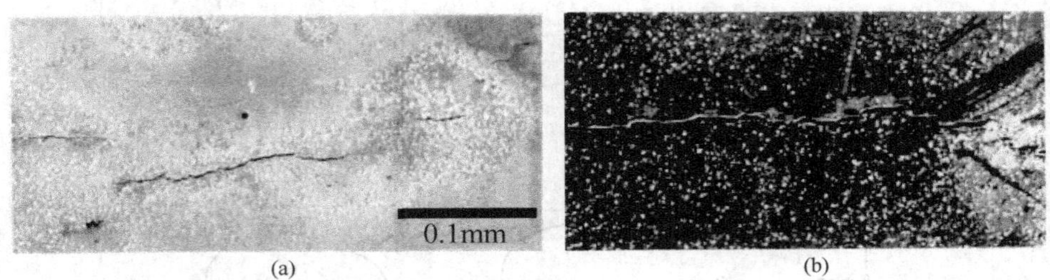

Figure 7.7 Stepwise cracking in a microalloyed steel after 24 h cathodic charging [From K. K.Chawla, J. M. Rigsbee, and J. D. Woodhouse, J. Mater. Sci., 21 (1986) 3777] .

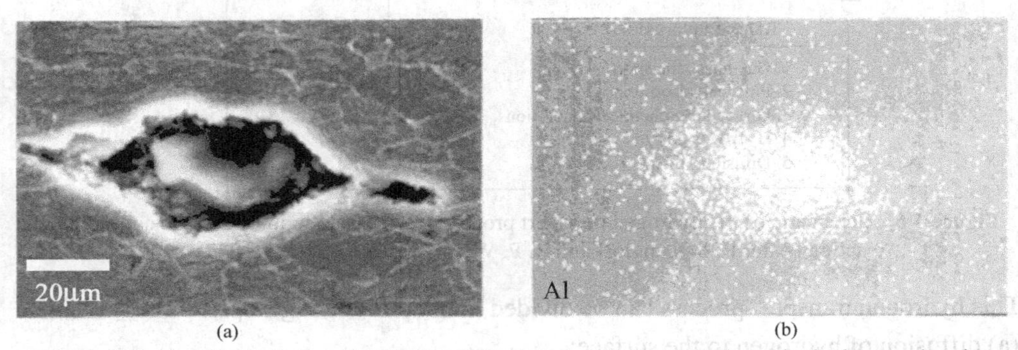

Figure 7.8 An aluminum-based inclusion in the interior of a void produced by hydrogen charging [From K. K. Chawla, J. M. Rigsbee, and J. D. Woodhouse, J. Mater. Sci., 21(1986) 3777].

The solubility of hydrogen is greatly influenced by the presence of lattice defects and impurities. For example, the solubility of hydrogen in a commercial steel at room temperature can be as much as one hundred percent greater than that in a clean and well-annealed steel. Thus, although the solubility of hydrogen in iron is small, a large amount of it can be trapped rather easily at various defect sites.

Gas or oil containing H_2S can lead to sulfide stress corrosion cracking or hydrogen induced blistering in steel. This form of HIC, also called blistering, is presumed to occur when hydrogen atoms generated in a wet, sour gas environment enter into the steel and precipitate at or around inclusions or other unfavorable microstructural sites.

In this regard, manganese sulfide inclusions, elongated in the rolling direction, are perhaps the worst culprits. Hydrogen atoms, generated at the surface, penetrate and diffuse into the steel. These atoms are trapped at matrix/inclusion interfaces and at ferrite/(pearlite + bainite+ martensite/austenite) interfaces. Here it is appropriate to point out an important microstructural feature of in rolled low carbon steels. It is tacitly assumed that the solute atoms in a solid solution are uniformly distributed in the matrix. More often than not, it is not the case. Indendritic segregation of solutes starts during the freezing of alloys. Specifically, in Mn-C steels interdendritic segregation of Mn, followed by rolling, can result in pronounced banding. Pearlite layers in the microstructure coincide with the Mn segregation. Such a microstructure consisting of alternate layers of ferrite and pearlite is very anisotropic and susceptible to hydrogen induced cracking. In quenched and tempered steels, even high Mn steels do not show such segregation; these steels have superior resistance to HIC.

Surface Energy

According to this theory, hydrogen is adsorbed on the free surfaces of a crack and reduces the surface energy. This results in a decrease in the work of fracture as per the Griffith criterion. This theory, however, would not explain the reversible degradation due to hydrogen.

Enhanced Plastic Flow

Beachem proposed a hydrogen assisted cracking model in which hydrogen enhances dislocation motion. The hydrogen diffuses in front of the crack tip, increases the mobility of dislocation there and causes, locally, an enhanced plasticity. Figure7.9 shows schematically this model. In the absence of hydrogen, a ductile metal fractures by microvoid coalescence within a large plastic zone at the crack tip. In the presence of hydrogen, however, locally plastic deformation becomes easier and crack growth occurs by severely localized deformation at the crack tip. This model has been supported experimentally by the work of Tabata and Birnbaum. They used an in-situ deformation stage in an environmental cell of a high voltage transmission electron microscope to investigate the effects of hydrogen on the behavior of dislocations in iron. It was observed that the introduction of hydrogen into the environmental cell increased the velocity of screw dislocations. This resulted in softening of the specimen in the early stages of deformation as the density of the mobile dislocations increased. In the later stages of deformation, this higher dislocation density may also contribute to work-hardening.

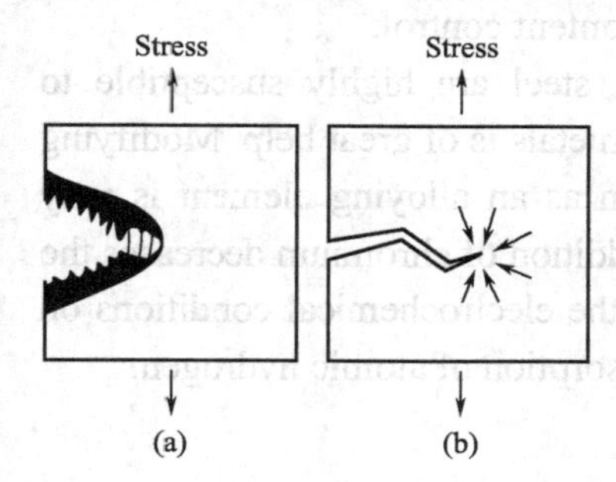

Figure 7.9 Schematic of crack growth in a high strength steel. (a) Without hydrogen, crack growth occurs by microvoid coalescence within a large plastic zone at the crack tip. (b) With hydrogen, plastic deformation becomes easy and crack growth occurs by severely localized deformation at the crack tip [After C. D. Beachem, Met. Trans., 3A (1972) 437].

These authors also studied the in-situ fracture behavior of iron of different purities in the presence of hydrogen gas and observed that the presence of hydrogen enhanced fracture. The main conclusions of this work of are:

(a) Basic fracture mechanisms in iron in vacuum and in hydrogen atmosphere are the same, but the morphology of fracture is very different.

(b) Hydrogen enhanced fracture is caused by the localization of plasticity and by the enhancement of dislocation motion and generation in the presence of hydrogen, as first suggested by Beachem.

Hydride Formation

Certain metals such as Ti, Zr, V, Nb, Ta, Mg, Al, etc. could suffer hydrogen degradation by diffusion of hydrogen and reaction with the metal to form a hydride at the crack tip. The hydride phase, being brittle, cracks easily on continued loading. Crack arrest occurs when the crack tip reaches the matrix phase. New hydride phase forms and the cycle is repeated as shown schematically in Figure 7.10. In pure iron, carbon and low alloy steels, a hydride phase is not formed or is unstable. This is attributed to the extremely low solubility of hydrogen in iron and steels. Other non-hydride forming systems include Mo, W, Cr, and their alloys.

Alleviating the Hydrogen Damage

While it is very difficult to provide simple recipes for alleviating the hydrogen damage in all the materials, we may list the following general guidelines as possible solutions:

(a) Avoid entry of hydrogen. This involves a control of the external environment, i.e., use of inhibitors or suitable alloying elements to protect the base metal surfaces against hydrogen ion discharge reaction.

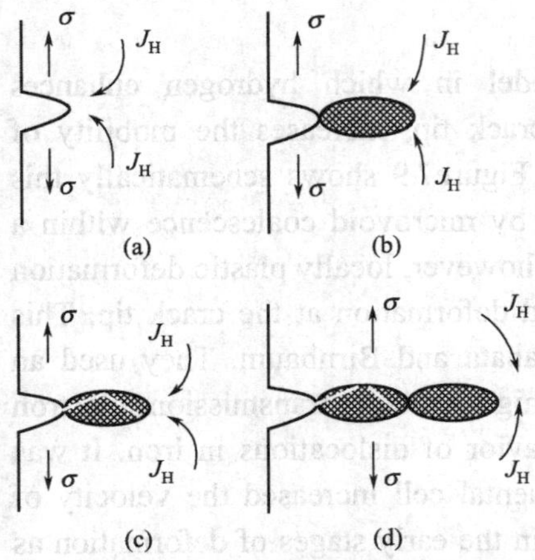

Figure 7.10 Hydrogen degradation due to a hydride formation. (a) Under stress, σ, hydrogen diffuses indicated by flux J_H , to the crack tip. (b) A hydride phase forms at the crack tip. (c) The brittle hydride phase cracks easily on continued loading. (d) New hydride phase forms and the cycle is repeated. [After H. K. Birnbaum, in Atomistics of Fracture (New York, Plenum, 1983), p. 733].

(b) Improve the material resistance to hydrogen damage. An effective way of doing this is to modify the morphology and/or decrease the number of inclusions. Lowering the sulfur content (maximum S about 0.010%) is a very important item in inclusion content control.

Because the elongated inclusions such as MnS stringers in steel are highly susceptible to hydrogen damage, inclusion shape control through use of rare earth metals is of great help. Modifying the alloy composition is yet another way. For example, chromium as an alloying element is very beneficial in steels. The reasons for this effect may be varied. The addition of chromium decreases the solubility of hydrogen in steels, perhaps, because chromium alters the electrochemical conditions on the surface of steel, enhances the oxidation of sulfur, or depresses adsorption of atomic hydrogen.

SUMMARY

Environmentally assisted fracture in metals can be classified under the following subheadings:

(1) stress corrosion cracking;

(2) hydrogen damage.

There are three regions in this curve for crack growth rate as function of the stress intensity factor under conditions of SCC. Region Ⅰ: In this region the crack velocity depends on the stress intensity factor. The threshold stress intensity, $K_{\mathrm{I\,scc}}$, below which the crack growth does not occur, is shown by a dashed line. Quite frequently, a true $K_{\mathrm{I\,scc}}$ does not exist. In such a case, we can define an operational $K_{\mathrm{I\,scc}}$ as that corresponding to a crack growth rate of $1\times10^{-9}\mathrm{ms}^{-1}$ or $1\times10^{-10}\mathrm{ms}^{-1}$. Such an arbitrary value can be used to rate different alloys. Region Ⅱ: The crack velocity in this region is independent of the stress intensity factor. The value at which this plateau region occurs is very specific to metal/environment combination and test conditions such as temperature. Region Ⅲ: In this region the crack velocity becomes very fast as the cracktip stress intensity factor approaches $K_{\mathrm{I\,c}}$. In this region, the crack velocity is mainly controlled by the stress intensity.

This hydrogen transport process can be divided into the following steps:

(a) diffusion of hydrogen to the surface;

(b) adsorption on the surface;

(c) dissociation in the surface adsorption layer;

(d) penetration through the surface;

(e) diffusion into the bulk of the metal.

IMPORTANT TERMS AND CONCEPTS

Electrochemical Corrosion	Galvanic Series	anode
reduction reaction	Passivity	Stress Corrosion Cracking (SCC)
Hydrogen Damage in Metals	Lattice Decohesion	Pressure theory
Hydride Formation	Surface Energy	Plastic Flow

REFERENCES

1 M. O. Speidel, Met. Trans., 6A (1975) 631

2 A. K. Vasudevan, P. R. Ziman, S. Jha, and T. H. Sanders, in Al-Li Alloys Ⅲ (London: The Institute of Metals, ,1986), p. 303

3 C. A. Wert, Phys. Rev., 79 (1959) 601

4 R. A. Oriani, in Fundamental Aspects of Stress Corrosion Cracking (Houston, TX: NACE, 1969), p. 32

5 J. C. M. Li, R. A. Oriani, and L. W. Darken, Z. Phys. Chem. 49 (1966) 271

6 R. P. Gangloff and R. P. Wei, Met. Trans., 8A (1977) 1043

7 A. R. Toriano, Trans. ASM, 52 (1960) 54

8 K. K. Chawla, J. M. Rigsbee, and J. D. Woodhouse, J. Mater. Sci., 21 (1986) 3777

9 D. D. J. Thomas and K. R. Doble, in Steels for Linepipe & Pipeline Fittings (London: The Metals Society1983), p. 22

10 T. Taira and Kobayashi, in Steels for Linepipe & Pipeline Fittings, (London: The Metals Society, 1983), p. 170

11 C. D. Beachem, Met. Trans., 3A (1972) 437

12 T. Tabata and H. K. Birnbaum, Scripta. Met., 17 (1983) 947

13 T. Tabata and H. K. Birnbaum, Scripta Met., 18 (1984) 231

14 H. K. Birnbaum, in Atomistics of Fracture, (New York: Plenum, 1983), p. 733

15 S. Al-Malaika, “Oxidative degradation and stabilisation of polymers,” Intl. Mater. Rev., 48 (2003)165

16 H. Arup and R. N. Parkins, (eds.), Stress Corrosion. Alphen aan den Rijn, The Netherlands: Sijthoff & Noordhoff, 1979

17 I. M. Bernstein and A. W. Thompson,eds. Hydrogen Effects in Metals. Warrendale, PA: TMS-AIME, 1981

18 M. R. Louthan, R. P. McNitt, and R. D. Sisson, eds. Environmental Degradation of Engineering Materials in Hydrogen. Blacksburg, VA: Virginia Tech Printing Dept., 1981

19 H. G. Nelson, ‘‘Hydrogen Embrittlement”, in Treatise on Materials Science and Technology, vol.25. New York: Academic Press, 1983, p. 275

20 D. Talbot and J. Talbot, Corrosion Science and Technology. Boca Raton, FL,:CRC Press, 1998

QUESTIONS AND PROBLEMS

7.1 Steel products are commonly protected by chromium or zinc coatings.
Based on the galvanic series, what difference would you expect in their ability to protect steel?

7.2 Explain why a small anode/cathode area ratio will result in a higher corrosion rate.

7.3 Alclad aluminum consists of a thin layer (5%~10% of total thickness) of one Al alloy metallurgically bonded to the core alloy. Generally, the cladding layer is anodic to the core. Why?

7.4 Tinplate (commonly used in the canning industry) is not plate or sheet of tin. It is actually a steel strip with a thin coating of tin. Discuss the pros and cons of using tin to protect steel.

7.5 Describe some methods of protecting the inside of a metallic pipe against chemical attack.

7.6 Describe some methods of protecting the inside of a metallic pipe against chemical attack.

7.7 A form of corrosion called pitting corrosion can occur in aluminum in fresh water. As the name suggests, pits form on the surface of aluminum in this type of corrosion. The pit depth, *d* follows a cube root relationship time, *t*:

$$d = A\, t^{1/3}.$$

Normally, a 5μm thick Al_2O_3 film forms on the surface of aluminum. If we double the thickness of the film, by what factor will the time to perforation increase?

7.8 Structural ceramic materials such as SiC, Si_3N_4, $MoSi_2$, etc. oxidize in the presence of oxygen at high temperatures. Give the oxidation reactions and indicate how the reaction products serve to protect these materials from further oxidation. Does it have deleterious effect on the high temperature capability of these materials?

7.9 A Ni-based superalloy has 0.2μm thick oxide layer. When placed in a burner rig to test for oxidation, it was observed to grow to 0.3μm in 1 h. If the superalloy follows a parabolic oxidation law ($x^2 = a + bt$, where x is the thickness, t is the time, and a and b are constants) what is the thickness after one week?

7.10 The stable, slow crack growth in a polymer in an aggressive environment can be represented by:

$$\frac{\mathrm{d}a}{\mathrm{d}t} = 0.03K_{\mathrm{I}}^{2} ,$$

Where a is the crack length in meters, t is the time in seconds, and K_{I} is the stress intensity factor in $\mathrm{MPa \cdot m^{1/2}}$. K_{Ic} for this polymer is 5 $\mathrm{MPa \cdot m^{1/2}}$. Calculate the time to failure under a constant applied stress of 50 MPa. Use $K_{\mathrm{I}} = \sigma\sqrt{\pi a}$.

Chapter 7

IMPORTANT TERMS AND CONCEPTS

Electrochemical Corrosion	电化学腐蚀	Cementite	渗碳体
Reduction Reaction	还原反应	Stress Corrosion Cracking , Scc	应力腐蚀断裂
Oxidation Reaction	氧化反应	Acid Solution	酸性溶液
Valence Electron	价电子	Galvanic Series	电动势顺序
Anode	阳极/正极	Passivity	钝性
Cathode	阴极/负极	Passivation	钝化
Surface Energy	表面能	Passivation Layer	钝化层
Crack Growth Velocity/ Crack Propagation Rate	裂纹扩展速率	Passivation Potential	钝化电势
Critical Stress Intensity Factor of Stress Corrosion, K_{Iscc}	应力腐蚀临界应力场强度因子，K_{Iscc}	Passive Region	钝化区
Threshold of Stress Corrosion, K_{Iscc}	应力腐蚀门槛值，K_{Iscc}	Transpassive Region	超钝化区
Cracktip Stress Intensity Factor, K_{I}	裂纹尖端应力场强度因子	Active Region	活化区
Ferrite	铁素体	Corrosion Current Density	腐蚀电流密度
Austenite	奥氏体	Hydrogen Embrittlement	氢脆
Pearlite	珠光体	Hydrogen Induced Cracking , Hic	氢脆断裂
Bainite	贝氏体	Indendritic Segregation	枝晶偏析
Martensite	马氏体	Plastic Flow	塑性流变/塑变

Chapter 8

Mechanical properties of ceramics

Learning Objectives

After careful study of this chapter you should be able to do the following:

1. Briefly explain why there is normally significant scatter in the fracture strength for identical specimens of the same ceramic material.

2. Compute the flexural strength of ceramic rod specimens that have been bent to fracture in three-point loading.

3. On the basis of slip considerations, explain why crystalline ceramic materials are normally brittle.

8.1 Introduction

Ceramic materials were discussed briefly in Chapter 1, which noted that they are inorganic and nonmetallic materials. Most ceramics are compounds between metallic and nonmetallic elements for which the interatomic bonds are either totally ionic, or predominantly ionic but having some covalent character. The term "ceramic" comes from the Greek word keramikos, which means "burnt stuff," indicating that desirable properties of these materials are normally achieved through a high-temperature heat treatment process called firing.

Up until the past 60 or so years, the most important materials in this class were termed the "traditional ceramics," those for which the primary raw material is clay; products considered to be traditional ceramics are china, porcelain, bricks, tiles, and, in addition, glasses and high-temperature ceramics. Of late, significant progress has been made in understanding the fundamental character of these materials and of the phenomena that occur in them that are responsible for their unique properties.

Consequently, a new generation of these materials has evolved, and the term "ceramic" has taken on a much broader meaning. To one degree or another, these new materials have a rather dramatic effect on our lives; electronic, computer, communication, aerospace, and a host of other industries rely on their use.

8.2 Stress-Strain Behavior

The stress-strain behavior of brittle ceramics is not usually ascertained by a tensile test as outlined in Section 8.2, for three reasons. First, it is difficult to prepare and test specimens having the required geometry. Second, it is difficult to grip brittle materials without fracturing them; and third, ceramics fail after only about 0.1% strain, which necessitates that tensile specimens be perfectly aligned to avoid the presence of bending stresses, which are not easily calculated. Therefore, a more suitable transverse bending test is most frequently employed, in which a rod specimen having either a circular or rectangular cross section is bent until fracture using a three- or four-point loading technique; the three-point loading scheme is illustrated in Figure 8.1. At the point of loading, the top surface of the specimen is placed in a state of compression, while the bottom surface is in tension. Stress is computed from the specimen thickness, the bending moment, and the moment of inertia of the cross section; these parameters are noted in Figure 8.1 for rectangular and circular cross sections. The maximum tensile stress (as determined using these stress expressions) exists at the bottom specimen surface directly below the point of load application. Since the tensile strengths of ceramics are about one-tenth of their compressive strengths, and since fracture occurs on the tensile specimen face, the flexure test is a reasonable substitute for the tensile test.

The stress at fracture using this flexure test is known as the flexural strength, modulus of rupture, fracture strength, or the bend strength, an important mechanical parameter for brittle ceramics. For a rectangular cross section, the flexural strength σ_{fs} is equal to

$$\sigma_{fs} = \frac{3F_f L}{2bd^2} \tag{8.1}$$

where F_f is the load at fracture, L is the distance between support points, and the other parameters are as indicated in Figure 8.1. When the cross section is circular, then

$$\sigma_{fs} = \frac{F_f L}{\pi R^3} \tag{8.2}$$

where R is the specimen radius.

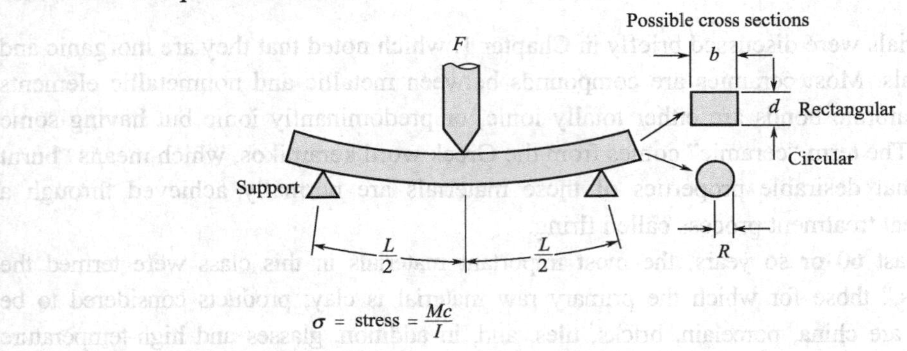

σ = stress = $\frac{Mc}{I}$

Where M = maximum bending moment

c = distance from center of specimen to outer fibers

I = moment of inertia of cross section

F = applied load

	M	c	I	σ
Rectangular	$\frac{FL}{4}$	$\frac{d}{2}$	$\frac{bd^3}{12}$	$\frac{3FL}{2bd^2}$
Circular	$\frac{FL}{4}$	R	$\frac{\pi R^4}{4}$	$\frac{FL}{\pi R^3}$

Figure 8.1 A three-point loading scheme for measuring the stress–strain behavior and flexural strength of brittle ceramics, including expressions for computing stress for rectangular and circular cross sections.

Characteristic flexural strength values for several ceramic materials are given in Table 8.1. Furthermore, σ_{fs} will depend on specimen size; as explained above, with increasing specimen volume (that is, specimen volume exposed to a tensile stress) there is an increase in the probability of the existence of a crack-producing flaw and, consequently, a decrease in flexural strength. In addition, the magnitude of flexural strength for a specific ceramic material will be greater than its fracture strength measured from a tensile test. This phenomenon may be explained by differences in specimen volume that are exposed to tensile stresses: the entirety of a tensile specimen is under tensile stress, whereas only some volume fraction of a flexural specimen is subjected to tensile stresses—those regions in the vicinity of the specimen surface opposite to the point of load application (Figure 8.1).

Table 8.1 Tabulation of Flexural Strength (Modulus of Rupture) **and Modulus of Elasticity for Ten Common Ceramic Materials**

Material	Flexural Strength		Modulus of Elasticity	
	MPa	ksi	GPa	1×10^6psi
Silicon nitride(Si_3N_4)	250~1000	35~145	304	44
Zirconia①(ZrO_2)	800~1500	115~215	205	30
Silicon carbide(SiC)	100~820	15~120	345	50
Aluminum oxide(Al_2O_3)	275~700	40~100	393	57
Glass-ceramic(Pyroceram)	247	36	120	17
Mullite($3Al_2O_3$-$2SiO_2$)	185	27	145	21
Spinel($MgAl_2O_4$)	110~245	16~35.5	260	38
Magnesium oxide(MgO)	105②	15②	225	33
Fused silica(SiO_2)	110	16	73	11
Soda-lime glass	69	10	69	10

① Partially stabilized with 3 mol% Y_2O_3.

② Sintered and containing approximately 5% porosity.

The elastic stress-strain behavior for ceramic materials using these flexure tests is similar to the tensile test results for metals: a linear relationship exists between stress and strain. Figure 8.2 compares the stress-strain behavior to fracture for aluminum oxide and glass. Again, the slope in the elastic region is the modulus of elasticity; the range of moduli of elasticity for ceramic materials is between about 70 and 500 GPa (10×10^6 and 70×10^6 psi), being slightly higher than for metals. Table 8.1 lists values for several ceramic materials. Also, from Figure 8.2 note that neither material experiences plastic deformation prior to fracture.

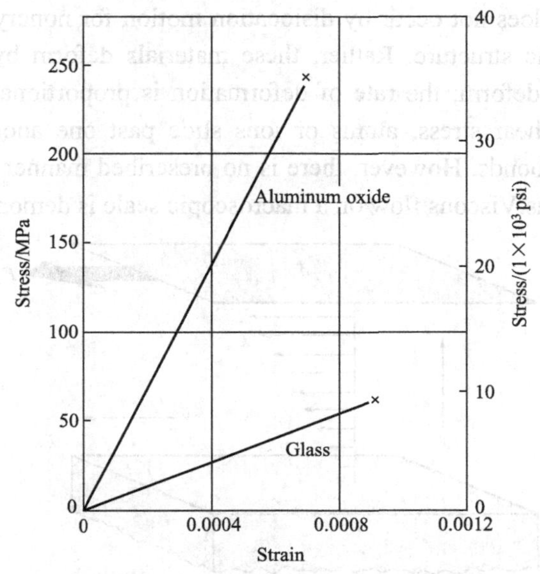

Figure 8.2 Typical stress-strain behavior to fracture for aluminum oxide and glass.

8.3 Mechanisms Of Plastic Deformation

Although at room temperature most ceramic materials suffer fracture before the onset of plastic deformation, a brief exploration into the possible mechanisms is worthwhile. Plastic deformation is different for crystalline and noncrystalline ceramics; however, each is discussed.

Although at room temperature most ceramic materials suffer fracture before the onset of plastic deformation, a brief exploration into the possible mechanisms is worthwhile. Plastic deformation is different for crystalline and noncrystalline ceramics; however, each is discussed.

8.3.1 Crystalline Ceramics

For crystalline ceramics, plastic deformation occurs, as with metals, by the motion of dislocations. One reason for the hardness and brittleness of these materials is the difficulty of slip (or dislocation motion). For crystalline ceramic materials for which the bonding is predominantly ionic, there are very few slip systems (crystallographic planes and directions within those planes) along which dislocations may move. This is a consequence of the electrically charged nature of the ions.For slip in some directions, ions of like charge are brought into close proximity to one another; because of electrostatic repulsion, this mode of slip is very restricted, to the extent that plastic deformation in ceramics is rarely measurable at room temperature. By way of contrast, in metals, since all atoms are electrically neutral, considerably more slip systems are operable and, consequently, dislocation motion is much more facile.

On the other hand, for ceramics in which the bonding is highly covalent, slip is also difficult and they are brittle for the following reasons:

(1) the covalent bonds are relatively strong,

(2) there are also limited numbers of slip systems,

(3) dislocation structures are complex.

8.3.2 Noncrystalline Ceramics

Plastic deformation does not occur by dislocation motion for noncrystalline ceramics because there is no regular atomic structure. Rather, these materials deform by viscous flow, the same manner in which liquids deform; the rate of deformation is proportional to the applied stress. In response to an applied shear stress, atoms or ions slide past one another by the breaking and reforming of interatomic bonds. However, there is no prescribed manner or direction in which this occurs, as with dislocations.Viscous flow on a macroscopic scale is demonstrated in Figure 8.3.

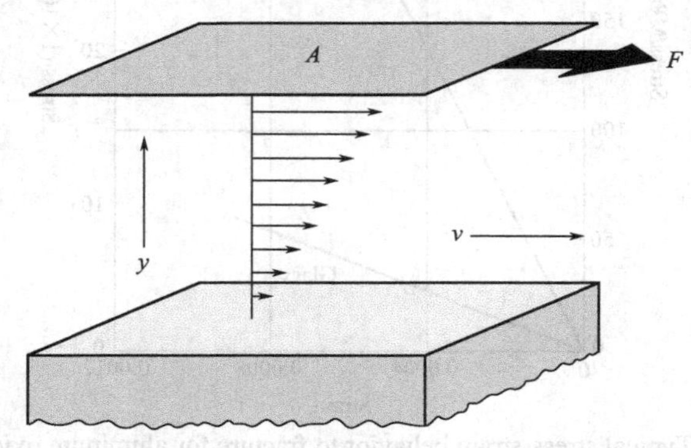

Figure 8.3 Representation of the viscous flow of a liquid or fluid glass in response to an applied shear force.

The characteristic property for viscous flow, viscosity, is a measure of a noncrystalline material's resistance to deformation. For viscous flow in a liquid that originates from shear stresses imposed by two flat and parallel plates, the viscosity η is the ratio of the applied shear stress τ and the change in velocity dv with distance dy in a direction perpendicular to and away from the plates, or

$$\eta = \frac{\tau}{\mathrm{d}v/\mathrm{d}y} = \frac{F/A}{\mathrm{d}v/\mathrm{d}y} \tag{8.3}$$

his scheme is represented in Figure 8.3.

The units for viscosity are poises (P) and pascal·seconds (Pa·s); 1P=1 dyne·s/cm^2, and 1 Pa·s or N·s/m^2. Liquids have relatively low viscosities; for example, the viscosity of water at room temperature is about 10^{-3} Pa • s. On the other hand, glasses have extremely large viscosities at ambient temperatures, which is accounted for by strong interatomic bonding. As the temperature is raised, the magnitude of the bonding is diminished, the sliding motion or flow of the atoms or ions is facilitated, and subsequently there is an attendant decrease in viscosity.

8.4 Brittle Fracture of Ceramics

Ceramic materials are somewhat limited in applicability by their mechanical properties, which in many respects are inferior to those of metals. The principal drawback is a disposition to catastrophic fracture in a brittle manner with very little energy absorption. At room temperature, both crystalline and noncrystalline ceramics almost always fracture before any plastic deformation can occur in response to an applied tensile load. The topics of brittle fracture and fracture mechanics, as discussed previously in Section 8.4, also relate to the fracture of ceramic materials; they will be reviewed briefly in this context.

The brittle fracture process consists of the formation and propagation of cracks through the cross section of material in a direction perpendicular to the applied load. Crack growth in crystalline ceramics may be either transgranular (i.e., through the grains) or intergranular (i.e., along grain boundaries); for transgranular fracture, cracks propagate along specific crystallographic (or cleavage) planes, planes of high atomic density.

The measured fracture strengths of ceramic materials are substantially lower than predicted by theory from interatomic bonding forces. This may be explained by very small and omnipresent flaws in the material that serve as stress raisers—points at which the magnitude of an applied tensile stress is amplified. The degree of stress amplification depends on crack length and tip radius of curvature according to Equation 3.1, being greatest for long and pointed flaws. These stress raisers may be minute surface or interior cracks (microcracks), internal pores, and grain corners, which are virtually impossible to eliminate or control. For example, even moisture and contaminants in the atmosphere can introduce surface cracks in freshly drawn glass fibers; these cracks deleteriously affect the strength. A stress concentration at a flaw tip can cause a crack to form, which may propagate until the eventual failure.The measure of a ceramic material's ability to resist fracture when a crack is present is specified in terms of fracture toughness. The plane strain fracture toughness K_{Ic}, as discussed in Section 4.5, is defined according to the expression

$$K_{\mathrm{Ic}} = Y\sigma\sqrt{\pi a} \tag{8.4}$$

where Y is a dimensionless parameter or function that depends on both specimen and crack geometries, is the applied stress, and a is the length of a surface crack or half of the length of an internal crack. Crack propagation will not occur as long as the righthand side of Equation 8.4 is less than the plane strain fracture toughness of the material. Plane strain fracture toughness values for ceramic materials are smaller than for metals; typically they are below $10\text{MPa}\sqrt{\text{m}}$ ($9\text{ksi}\sqrt{\text{in}}$).

Under some circumstances, fracture of ceramic materials will occur by the slow propagation of cracks, when stresses are static in nature, and the right-hand side of Equation 8.4 is less than K_{Ic}. This phenomenon is called static fatigue, or delayed fracture; use of the term "fatigue" is somewhat misleading inasmuch as fracture may occur in the absence of cyclic stresses (metal fatigue was discussed in Chapter 3). It has been observed that this type of fracture is especially sensitive to environmental conditions, specifically when moisture is present in the atmosphere. Relative to mechanism, a stress-corrosion process probably occurs at the crack tips. That is, the combination of an applied tensile stress and atmospheric moisture at crack tips causes ionic bonds to rupture; this leads to a sharpening and lengthening of the cracks until, ultimately, one crack grows to a size capable of rapid propagation according to Equation 3.3. Furthermore, the duration of stress application preceding fracture diminishes with increasing stress. Consequently, when specifying the static fatigue strength, the time of stress application should also be stipulated. Silicate glasses are especially susceptible to this type of fracture; it has also been observed in other ceramic materials to include porcelain, portland cement, high-alumina ceramics, barium titanate, and silicon nitride.

There is usually considerable variation and scatter in the fracture strength for many specimens of a specific brittle ceramic material. A distribution of fracture strengths for a silicon nitride material is shown in Figure 8.4.This phenomenon may be explained by the dependence of fracture strength on the probability of the existence of a flaw that is capable of initiating a crack. This probability varies from specimen to specimen of the same material and depends on fabrication technique and any subsequent treatment. Specimen size or volume also influences fracture strength; the larger the specimen, the greater is this flaw existence probability, and the lower the fracture strength.

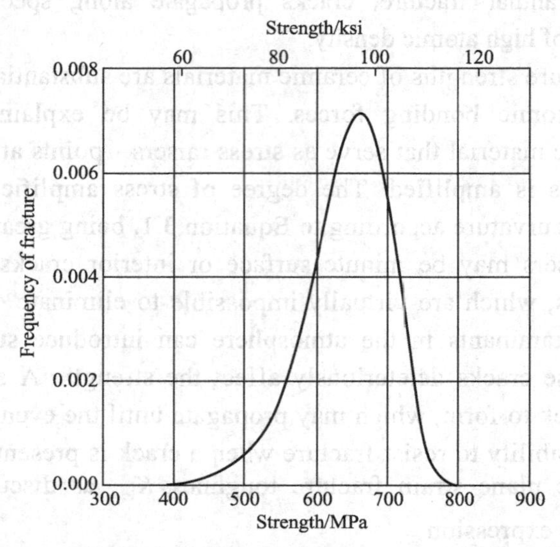

Figure 8.4 The frequency distribution of observed fracture strengths for a silicon nitride material.

For compressive stresses, there is no stress amplification associated with any existent flaws. For this reason, brittle ceramics display much higher strengths in compression than in tension (on the order of a factor of 10), and they are generally utilized when load conditions are compressive. Also, the fracture strength of a brittle ceramic may be enhanced dramatically by imposing residual compressive stresses at its surface. One way this may be accomplished is by thermal tempering.

Statistical theories have been developed that in conjunction with experimental data are used to determine the risk of fracture for a given material; a discussion of these is beyond the scope of the present treatment. However, due to the dispersion in the measured fracture strengths of brittle ceramic materials, average values and factors of safety are not normally employed for design purposes.

Fractography of Ceramics

It is sometimes necessary to acquire information regarding the cause of a ceramic fracture so that measures may be taken to reduce the likelihood of future incidents. A failure analysis normally focuses on determination of the location, type, and source of the crack-initiating flaw. A fractographic study (Section 3.2) is normally a part of such an analysis, which involves examining the path of crack propagation as well as microscopic features of the fracture surface. It is often possible to conduct an investigation of this type using simple and inexpensive equipment—for example, a magnifying glass, and/or a low-power stereo binocular optical microscope in conjunction with a light source. When higher magnifications are required the scanning electron microscope is utilized.

After nucleation, and during propagation, a crack accelerates until a critical (or terminal) velocity is achieved; for glass, this critical value is approximately one-half of the speed of sound. Upon reaching this critical velocity, a crack may branch (or bifurcate), a process that may be successively repeated until a family of cracks is produced. Typical crack configurations for four common loading schemes are shown in Figure 8.5. The site of nucleation can often be traced back to the point where a set of cracks converges or comes together. Furthermore, rate of crack acceleration increases with increasing stress level; correspondingly degree of branching also increases with rising stress. For example, from experience we know that when a large rock strikes (and probably breaks) a window, more crack branching results [i.e., more and smaller cracks form (or more broken fragments are produced)] than for a small pebble impact.

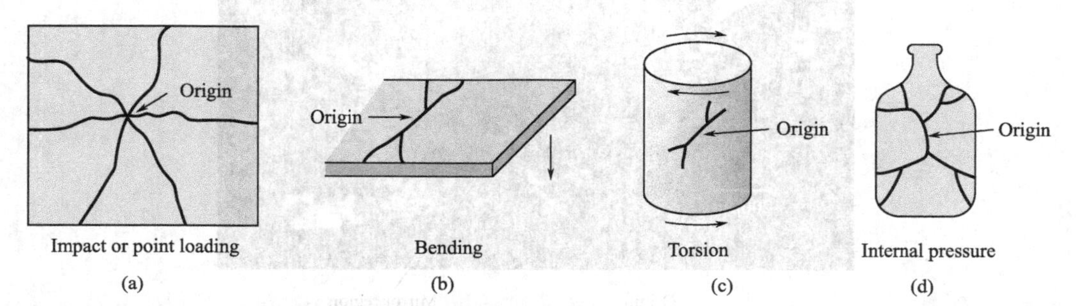

Figure 8.5 For brittle ceramic materials, schematic representations of crack origins and configurations that result from (a) impact (point contact) loading, (b) bending, (c) torsional loading, and (d) internal pressure (From D.W. Richerson, Modern Ceramic Engineering, 2nd edition, Marcel Dekker, Inc., New York, 1992. Reprinted from Modern Ceramic Engineering, 2nd edition, p. 681, by courtesy of Marcel Dekker, Inc) .

During propagation, a crack interacts with the microstructure of the material, with the stress, as well as with elastic waves that are generated; these interactions produce distinctive features on the fracture surface. Furthermore, these features provide important information on where the crack initiated, and the source of the crackproducing defect. In addition, measurement of the approximate fracture-producing stress may be useful; stress magnitude is indicative of whether the ceramic piece was excessively weak, or the in-service stress was greater than anticipated.

Several microscopic features normally found on the crack surfaces of failed ceramic pieces are shown in the schematic diagram of Figure 8.6 and also the photomicrograph in Figure 8.7. The crack surface that formed during the initial acceleration stage of propagation is flat and smooth, and appropriately termed the mirror region (Figure 8.6). For glass fractures, this mirror region is extremely flat and highly reflective; on the other hand, for polycrystalline ceramics, the flat mirror surfaces are rougher and have a granular texture. The outer perimeter of the mirror region is roughly circular, with the crack origin at its center.

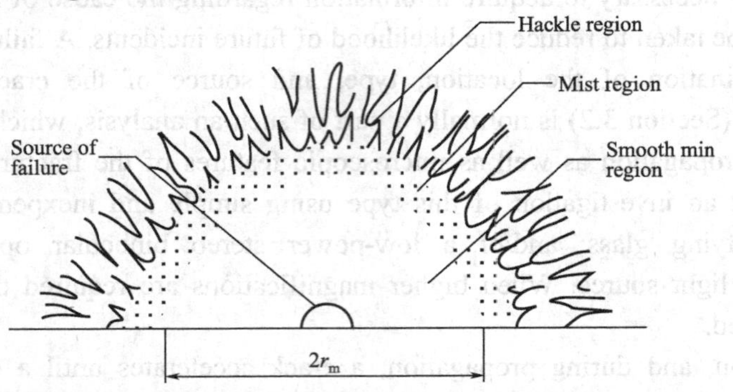

Figure 8.6 Schematic diagram that shows typical features observed on the fracture surface of a brittle ceramic [Adapted from J. J. Mecholsky, R.W. Rice, and S.W. Freiman, "Prediction of Fracture Energy and Flaw Size in Glasses from Measurements of Mirror Size," J. Am. Ceram. Soc., 57 [10] 440 (1974)].

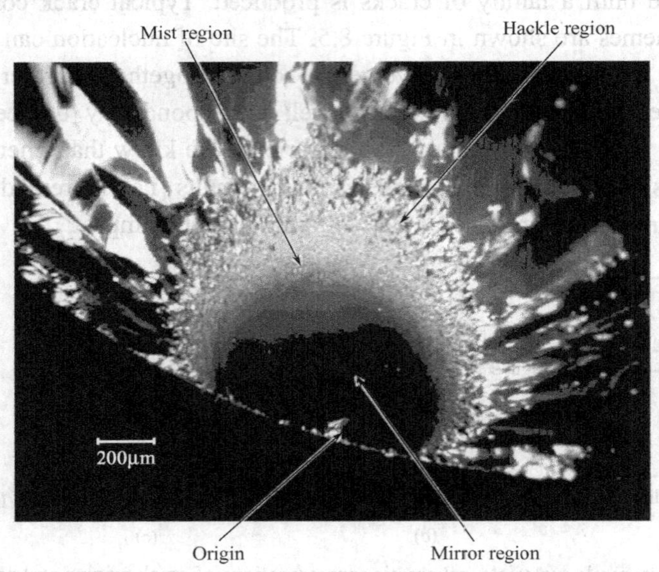

Figure 8.7 Photomicrograph of the fracture surface of a 6 mm-diameter fused silica rod that was fractured in four-point bending. Features typical of this kind of fracture are noted—i.e., the origin, as well as mirror, mist, and hackle regions 500× (Courtesy of George Quinn, National Institute of Standards and Technology, Gaithersburg, MD).

Upon reaching its critical velocity, the crack begins to branch—that is, the crack surface changes propagation direction. At this time there is a roughening of the crack interface on a microscopic scale, and the formation of two more surface features—mist and hackle; these are also noted in Figure 8.6 and Figure 8.7. The mist is a faint annular region just outside the mirror; it is often not discernible for polycrystalline ceramic pieces. And beyond the mist is the hackle, which has an even rougher texture. The hackle is composed of a set of striations or lines that radiate away from the crack source in the direction of crack propagation; furthermore, they intersect near the crack initiation site, and may be used to pinpoint its location. Qualitative information regarding the magnitude of the fracture-producing stress is available from measurement of the mirror radius (r_m in Figure 8.6). This radius is a function of the acceleration rate of a newly formed crack—that is, the greater this acceleration rate, the sooner the crack reaches its critical velocity and the smaller the mirror radius. Furthermore, the acceleration rate increases with stress level. Thus, as fracture stress level increases, the mirror radius decreases; experimentally it has been observed that

$$\sigma_f \propto \frac{1}{r_m^{0.5}} \tag{8.5}$$

Here σ_f is the stress level at which fracture occurred.

Elastic (sonic) waves are generated also during a fracture event, and the locus of intersections of these waves with a propagating crack front give rise to another type of surface feature known as a Wallner line. Wallner lines are arc shaped, and they provide information regarding stress distributions and directions of crack propagation.

8.5 Miscellaneous Mechanical Considerations

8.5.1 Influence of Porosity

For some ceramic fabrication techniques, the precursor material is in the form of a powder. Subsequent to compaction or forming of these powder particles into the desired shape, pores or void spaces will exist between the powder particles. During the ensuing heat treatment, much of this porosity will be eliminated; however, it is often the case that this pore elimination process is incomplete and some residual porosity will remain. Any residual porosity will have a deleterious influence on both the elastic properties and strength. For example, it has been observed for some ceramic materials that the magnitude of the modulus of elasticity E decreases with volume fraction porosity P according to

$$E = E_0(1-1.9P+0.9P^2) \tag{8.6}$$

where E_0 is the modulus of elasticity of the nonporous material. The influence of volume fraction porosity on the modulus of elasticity for aluminum oxide is shown in Figure 8.8; the curve represented in the figure is according to Equation 8.6.

Porosity is deleterious to the flexural strength for two reasons:

(1) pores reduce the cross-sectional area across which a load is applied,

(2) they also act as stress concentrators—for an isolated spherical pore, an applied tensile stress is amplified by a factor of 2. The influence of porosity on strength is rather dramatic; for example, it is not uncommon that 10 vol% porosity will decrease the flexural strength by 50% from the measured value for the nonporous material. The degree of the influence of pore volume on flexural strength is demonstrated in Figure 8.9, again for aluminum oxide. Experimentally it has been shown that the flexural strength decreases exponentially with volume fraction porosity (P) as

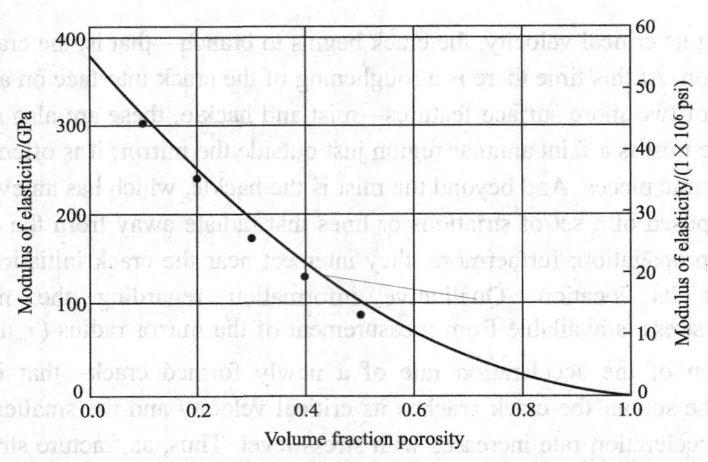

Figure 8.8 The influence of porosity on the modulus of elasticity for aluminum oxide at room temperature. The curve drawn is according to Equation 8.6 (From R. L. Coble and W. D. Kingery, "Effect of Porosity on Physical Properties of Sintered Alumina," J. Am.Ceram. Soc., **39,** 11, Nov. 1956, p. 381).

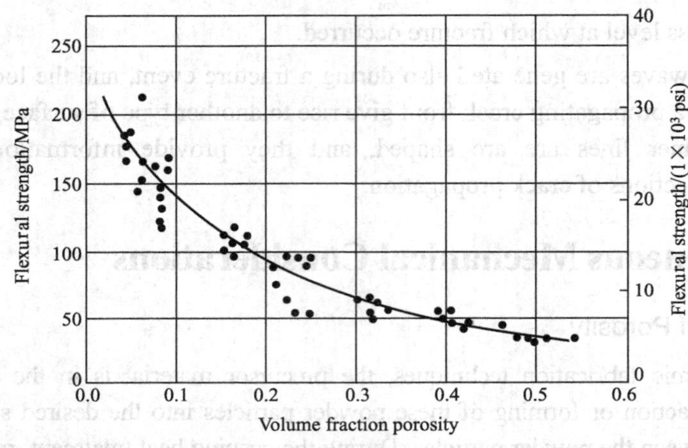

Figure 8.9 The influence of porosity on the flexural strength for aluminum oxide at room temperature (From R. L. Coble and W. D. Kingery, "Effect of Porosity on Physical Properties of Sintered Alumina," J. Am. Ceram. Soc., **39,** 11, Nov. 1956, p. 382).

$$\sigma_{fs} = \sigma_0 \exp(-np) \tag{8.7}$$

In this expression σ_0 and n are experimental constants.

8.5.2 Hardness

One beneficial mechanical property of ceramics is their hardness, which is often utilized when an abrasive or grinding action is required; in fact, the hardest known materials are ceramics. A listing of a number of different ceramic materials according to Knoop hardness is contained in Table 8.2. Only ceramics having Knoop hardnesses of about 1000 or greater are utilized for their abrasive characteristics.

Table 8.2 Approximate Knoop Hardness (100 g load) **for Seven Ceramic Materials**

Material	Approximate Knoop Hardness	Material	Approximate Knoop Hardness
Diamond(carbon)	7000	Aluminum oxide(Al_2O_3)	2100
Boron carbide(B_4C)	2800	Quartz(SiO_2)	800
Silicon carbide(SiC)	2500	Glass	550
Tungsten carbide(WC)	2100		

8.5.3 Creep

Often ceramic materials experience creep deformation as a result of exposure to stresses (usually compressive) at elevated temperatures. In general, the time-deformation creep behavior of ceramics is similar to that of metals; however, creep occurs at higher temperatures in ceramics. High-temperature compressive creep tests are conducted on ceramic materials to ascertain creep deformation as a function of temperature and stress level.

SUMMARY

Stress–Strain Behavior

At room temperature, virtually all ceramics are brittle. Microcracks, the presence of which is very difficult to control, result in amplification of applied tensile stresses and account for relatively low fracture strengths (flexural strengths). This amplification does not occur with compressive loads, and, consequently, ceramics are stronger in compression.

Mechanisms of Plastic Deformation

Any plastic deformation of crystalline ceramics is a result of dislocation motion; the brittleness of these materials is explained, in part, by the limited number of operable slip systems. The mode of plastic deformation for noncrystalline materials is by viscous flow; a material's resistance to deformation is expressed as viscosity. At room temperature, the viscosities of many noncrystalline ceramics are extremely high.

Brittle Fracture of Ceramics

Fractographic analysis of the fracture surface of a ceramic material may reveal the location and source of the crack-producing flaw, as well as the magnitude of the fracture stress. Representative strengths of ceramic materials are determined by performing transverse bending tests to fracture.

Miscellaneous Mechanical Considerations

Many ceramic bodies contain residual porosity, which is deleterious to both their moduli of elasticity and fracture strengths. In addition to their inherent brittleness, ceramic materials are distinctively hard. Also, since these materials are frequently utilized at elevated temperatures and under applied loads, creep characteristics are important.

REFERENCES

1. Barsoum, M. W., Fundamentals of Ceramics, McGraw-Hill, New York, 1997
2. Bergeron, C. G. and S. H. Risbud, Introduction to Phase Equilibria in Ceramics, American Ceramic Society, Columbus, OH, 1984
3. Bowen, H. K., "Advanced Ceramics," Scientific American, Vol. 255, No. 4, October 1986, p. 168～176
4. Chiang, Y. M., D. P. Birnie, III, and W. D. Kingery, Physical Ceramics: Principles for Ceramic Science and Engineering, Wiley, New York, 1997
5. Curl, R. F. and R. E. Smalley, "Fullerenes," Scientific American, Vol. 265, No. 4, October 1991,p. 54～63
6. Davidge, R.W., Mechanical Behaviour of Ceramics, Cambridge University Press, Cambridge, 1979. Reprinted by TechBooks, Marietta, OH, 1988
7. Doremus, R. H., Glass Science, 2nd edition,Wiley, New York, 1994

QUESTIONS AND PROBLEMS

8.1 A three-point bending test is performed on a spinel ($MgAl_2O_4$) specimen having a rectangular cross section of height d 3.8 mm (0.15in) and width b 9 mm (0.35in); the distance between support points is 25mm (1.0in).

(a) Compute the flexural strength if the load at fracture is 350N (80lb_f).

(b) The point of maximum deflection Δy occurs at the center of the specimen and is described by $\Delta y = \dfrac{FL^3}{48EI}$, where E is the modulus of elasticity and I is the cross-sectional moment of inertia. Compute at a load of 310N (70lb$_f$).

8.2 A circular specimen of MgO is loaded using a three-point bending mode. Compute the minimum possible radius of the specimen without fracture, given that the applied load is 5560 N (1250lb$_f$), the flexural strength is 105MPa (15000psi), and the separation between load points is 45 mm (1.75in).

8.3 A three-point bending test was performed on an aluminum oxide specimen having a circular cross section of radius 5.0mm (0.20in); the specimen fractured at a load of 3000 N (675lb$_f$) when the distance between the support points was 40 mm (1.6in). Another test is to be performed on a specimen of this same material, but one that has a square cross section of 15 mm (0.6in) length on each edge. At what load would you expect this specimen to fracture if the support point separation is maintained at 40 mm (1.6in)?

8.4 (a) A three-point transverse bending test is conducted on a cylindrical specimen of aluminum oxide having a reported flexural strength of 300MPa (43500psi). If the specimen radius is 5.0mm (0.20in) and the support point separation distance is 15.0mm (0.61in), predict whether or not you would expect the specimen to fracture when a load of 7500N (1690lb$_f$) is applied? Justify your prediction.

(b) Would you be 100% certain of the prediction in part (a)? Why or why not?

8.5 Cite one reason why ceramic materials are, in general, harder yet more brittle than metals.

8.6 The modulus of elasticity for spinel ($MgAl_2O_4$) having 5 vol% porosity is 240GPa (35×10^6psi).

(a) Compute the modulus of elasticity for the nonporous material.

(b) Compute the modulus of elasticity for 15 vol% porosity.

8.7 The modulus of elasticity for titanium carbide (TiC) having 5 vol% porosity is 310GPa (45×10^6psi).

(a) Compute the modulus of elasticity for the nonporous material.

(b) At what volume percent porosity will the modulus of elasticity be 240GPa (35×10^6psi)?

Chapter 8

IMPORTANT TERMS AND CONCEPTS

Ceramics	陶瓷	Static Fatigue/ Delayed Fracture	静态疲劳
Crystalline Ceramics	晶体陶瓷	Hackle Region	锯齿形区
Noncrystalline Ceramics	非晶陶瓷	Mist Region	细雾区
Flexural Strength/ Bend Strength	抗弯强度	Smooth Mirror Region	匀镜区
Three-point Bend	三点弯曲	Viscous Flow	黏性流
Four-point Bend	四点弯曲	Porosity	气孔/松孔/孔隙

Chapter 9

Mechanical properties of Polymers

Learning ObjeSctives

After careful study of this chapter you should be able to do the following:

1. Make schematic plots of the three characteristic stress–strain behaviors observed for polymeric materials.

2. Describe/sketch the various stages in the elastic and plastic deformations of a semicrystalline (spherulitic) polymer.

3. Discuss the influence of the following factors on polymer tensile modulus and/or strength:

(a) molecular weight,

(b) degree of crystallinity,

(c) predeformation,

(d) heat treating of undeformed materials.

4. Describe the molecular mechanism by which elastomeric polymers deform elastically.

9.1 Introduction

Polymers are used in a wide variety of applications from construction materials to microelectronics processing. Thus, most engineers will be required to work with polymers at some point in their careers. Understanding the mechanisms by which polymers elastically and plastically deform allows one to alter and control their moduli of elasticity and strengths.

9.2 Stress-Strain Behavior

The mechanical properties of polymers are specified with many of the same parameters that are used for metals—that is, modulus of elasticity, and yield and tensile strengths. For many polymeric materials, the simple stress-strain test is employed for the characterization of some of these mechanical parameters. The mechanical characteristics of polymers, for the most part, are highly sensitive to the rate of deformation (strain rate), the temperature, and the chemical nature of the environment (the presence of water, oxygen, organic solvents, etc.). Some modifications of the testing techniques and specimen configurations used for metals are necessary with polymers, especially for the highly elastic materials, such as rubbers.

Three typically different types of stress-strain behavior are found for polymeric materials, as represented in Figure 9.1. Curve A illustrates the stress–strain character for a brittle polymer, inasmuch as it fractures while deforming elastically. The behavior for a plastic material, curve B, is similar to that for many metallic materials; the initial deformation is elastic, which is followed by yielding and a region of plastic deformation. Finally, the deformation displayed by curve C is totally elastic; this rubber-like elasticity (large recoverable strains produced at low stress levels) is displayed by a class of polymers termed the elastomers.

Modulus of elasticity (termed tensile modulus or sometimes just modulus for polymers) and ductility in percent elongation are determined for polymers in the same manner as for metals. For plastic polymers (curve *B*, Figure 9.1), the yield point is taken as a maximum on the curve, which occurs just beyond the termination of the linear-elastic region (Figure 9.2). The stress at this maximum is the yield strength (σ_y). Furthermore, tensile strength (*TS*) corresponds to the stress at which fracture occurs (Figure 9.2); *TS* may be greater than or less than σ_y. Strength, for these plastic polymers, is normally taken as tensile strength. Table 9.1 gives these mechanical properties for several polymeric materials.

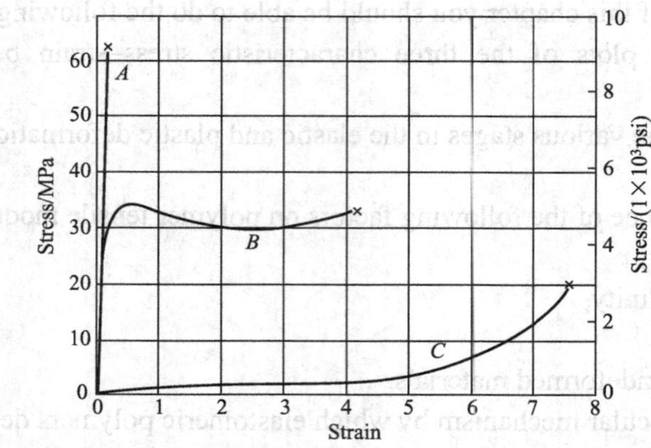

Figure 9.1 The stress-strain behavior for brittle (curve *A*), plastic (curve *B*), and highly elastic (elastomeric,curve *C*) polymers.

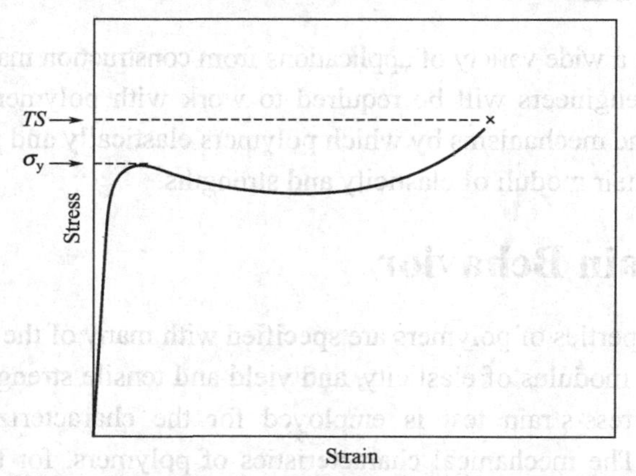

Figure 9.2 Schematic stress-strain curve for a plastic polymer showing how yield and tensile strengths are determined.

Polymers are, in many respects, mechanically dissimilar to metals. For example, the modulus for highly elastic polymeric materials may be as low as 7MPa(1×10^3 psi) but may run as high as 4GPa (0.6×10^6psi) for some of the very stiff polymers; modulus values for metals are much larger and range between 48 and 410 GPa(7×10^6 to 60×10^6 psi). Maximum tensile strengths for polymers are about 100MPa (15000psi)—for some metal alloys 4100MPa (600000psi).And, whereas metals rarely elongate plastically to more than 100%, some highly elastic polymers may experience elongations to greater than 1000%.

Table 9.1 Room-Temperature Mechanical Characteristics of Some of the More Common Polymers

Material	Specific Gravity	Tensile Modulus /GPa (ksi)	Tensile Strength /MPa (ksi)	Yield Strength /MPa (ksi)	Elongation at Break/%
Polyethylene(low density)	0.917~0.932	0.17~0.28 (25~41)	8.3~31.4 (1.2~4.55)	9.0~14.5 (1.3~2.1)	100~650
Polyethylene(high density)	0.952~0.965	1.60~1.09 (155~158)	22.1~31.0 (3.2~4.5)	26.2~33.1 (3.8~4.8)	10~1200
Poly(vinyl chloride)	1.30~1.58	2.4~4.1 (350~600)	40.7~51.7 (5.9~7.5)	40.7~44.8 (5.9~6.5)	40~80
Polytetrafluoroethylene	2.14~2.20	0.40~0.55 (58~80)	20.7~34.5 (3.0~5.0)	—	200~400
Polypropylene	0.90~0.91	1.14~1.55 (165~225)	31~41.4 (4.5~6.0)	31.0~37.2 (4.5~5.4)	100~600
Polystyrene	1.04~1.05	2.28~3.28 (330~475)	35.9~51.7 (5.2~7.5)	—	1.2~2.5
Poly(methyl methacrylate)	1.17~1.20	2.24~3.24 (325~470)	48.3~72.4 (7.0~10.5)	53.8~73.1 (7.8~10.6)	2.0~5.5
Phenol-formaldehyde	1.24~1.32	2.76~4.83 (400~700)	34.5~62.1 (5.0~9.0)	—	1.5~2.0
Nylon 6,6	1.13~1.15	1.58~3.80 (230~550)	75.9~94.5 (11.0~13.7)	44.8~82.8 (6.5~12)	15~300
Polyester(PET)	1.29~1.40	2.8~4.1 (400~600)	48.3~72.4 (7.0~10.5)	59.3 (8.6)	30~300
Polycarbonate	1.20	2.38 (345)	62.8~72.4 (9.1~10.5)	62.1 (9.0)	110~150

Source: *Modern Plastics Encyclopedia* '96. Copyright 1995, The McGraw-Hill Companies. Reprinted with permission.

In addition, the mechanical characteristics of polymers are much more sensitive to temperature changes near room temperature. Consider the stress-strain behavior for poly (methyl methacrylate) (Plexiglas) at several temperatures between 4ºC and 60ºC (40ºF and 140ºF, Figure 9.3). It should be noted that increasing the temperature produces:

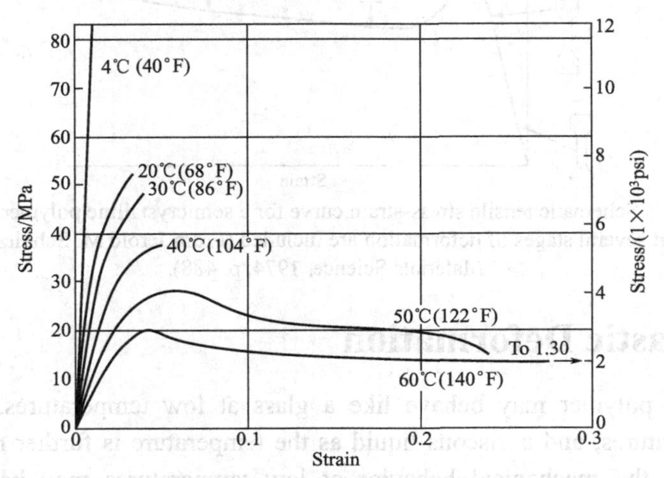

Figure 9.3 The influence of temperature on the stress-strain characteristics of poly (methyl methacrylate) (From T. S. Carswell and H. K. Nason, "Effect of Environmental Conditions on the Mechanical Properties of Organic Plastics," Symposium on Plastics, American Society for Testing and Materials, Philadelphia, 1944).

(1) a decrease in elastic modulus,

(2) a reduction in tensile strength,

(3) an enhancement of ductility—at 4ºC(40ºF) the material is totally brittle, while there is considerable plastic deformation at both 50ºC and 60ºC (122ºF and 140ºF).

The influence of strain rate on the mechanical behavior may also be important. In general, decreasing the rate of deformation has the same influence on the stress-strain characteristics as increasing the temperature; that is, the material becomes softer and more ductile.

9.3 Macroscopic Deformation

Some aspects of the macroscopic deformation of semicrystalline polymers deserve our attention. The tensile stress-strain curve for a semicrystalline material, which was initially undeformed, is shown in Figure 9.4; also included in the figure are schematic representations of the specimen profiles at various stages of deformation. Both upper and lower yield points are evident on the curve, which are followed by a near horizontal region. At the upper yield point, a small neck forms within the gauge section of the specimen. Within this neck, the chains become oriented (i.e., chain axes become aligned parallel to the elongation direction, a condition that is represented schematically in Figure 9.14), which leads to localized strengthening. Consequently, there is a resistance to continued deformation at this point, and specimen elongation proceeds by the propagation of this neck region along the gauge length; the chain orientation phenomenon (Figure 9.14) accompanies this neck extension. This tensile behavior may be contrasted to that found for ductile metals, wherein once a neck has formed, all subsequent deformation is confined to within the neck region.

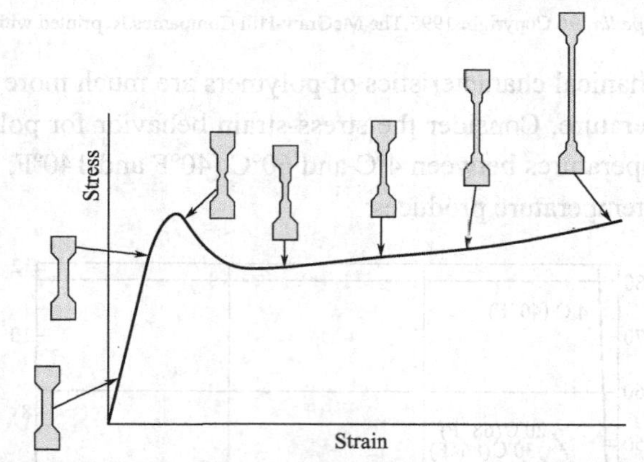

Figure 9.4 Schematic tensile stress-strain curve for a semicrystalline polymer. Specimen contours at several stages of deformation are included (From Jerold M. Schultz, Polymer Materials Science, 1974, p. 488).

9.4 Viscoelastic Deformation

An amorphous polymer may behave like a glass at low temperatures, a rubbery solid at intermediate temperatures, and a viscous liquid as the temperature is further raised. For relatively small deformations, the mechanical behavior at low temperatures may be elastic; that is, in conformity to Hooke's law, $\sigma=E\varepsilon$. At the highest temperatures, viscous or liquidlike behavior prevails. For intermediate temperatures the polymer is a rubbery solid that exhibits the combined mechanical characteristics of these two extremes; the condition is termed viscoelasticity.

The deformation of an amorphous material does not involve atomic displacements on specific crystallographic planes, as is the case in crystalline metals. Rather, a continuous displacement of atoms or molecules takes place with time at a constant load. This flow mechanism of noncrystalline materials is associated with the diffusion of atoms or molecules within the material; that is, it is a thermally activated process and is thus described by an Arrhenius-type equation. Of course, at sufficiently high temperatures, where diffusion becomes important, crystalline as well as amorphous materials show a large amount of thermally activated plastic flow. Liquids and even fluids in general show a characteristic resistance to flow called viscosity. The viscosity of a fluid results in a frictional energy loss, which appears as heat. The more viscous a fluid, the higher is the frictional energy loss.

Over a range of temperatures, the viscosity η can be described by the Arrhenius-type relationship

$$1/\eta = A\exp(-Q/RT) \tag{9.1}$$

where Q represents the activation energy for the atomic or molecular process responsible for the viscosity, R is the universal gas constant, and T is the temperature in kelvin. The S. I. units of the viscosity η are $\mathrm{N \cdot m^{-2}\, s}$ or Pa· s. Another common unit of viscosity is poise, P; 1P = 0.1Pa·s.

A purely viscous material shows stress proportional to strain rate. Thus, if we apply a shear stress τ to a glassy solid above its glass transition temperature, then we can write, for the rate of shear deformation,

$$\dot{\gamma} = \frac{\mathrm{d}\gamma}{\mathrm{d}t} = \frac{\tau}{\eta} = \phi\tau \tag{9.2}$$

where ϕ is the *fluidity* (the reciprocal of viscosity) of the material.

Equation 9.2can be written as

$$\tau = \eta\dot{\gamma} \tag{9.3}$$

If the viscosity of a material does not change with the strain rate (i.e., if the stress is linearly proportional to the strain rate), then we call the viscosity a Newtonian viscosity and such a material a Newtonian material. Figure 9.5 shows a Newtonian (or linear) response curve. If the stress is not directly proportional to the strain rate, we have a non-Newtonian response, which can be written as

$$\tau = \eta\dot{\gamma} \tag{9.4}$$

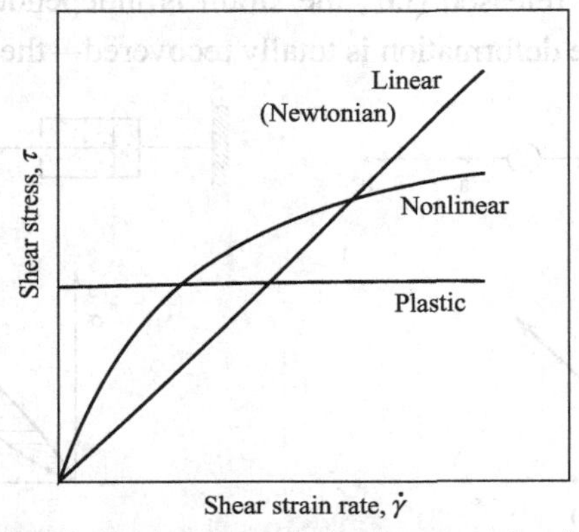

Figure. 9.5 Linear or Newtonian response (stress proportional to deformation rate), nonlinear response, and plastic response (stress independent of deformation rate).

This is shown by the curve marked "nonlinear" in the figure. If the stress is independent of the strain rate, we have a plastic material. A special case is that of a material whose viscosity decreases when subjected to high strain rates. Such a material is called a thixotropic material, a good example of which is a latex paint. When we apply the paint to a vertical wall, it does not sag, because its viscosity is very high on the wall. However, we can stir and brush the paint easily because its viscosity decreases when subjected to shear stress in the stirring action.

Polymers, polymer solutions and dispersions, metals at very high temperatures, and amorphous materials (organic and inorganic) show viscoelastic behavior—that is, characteristics intermediate between perfectly elastic and perfectly viscous behavior. Commercial silica-based glasses have a high proportion of additives: about 30% in soda—lime glass and 20% in high-temperature glasses such as Pyrex. The main purpose of the additives is to lower the viscosity by breaking up the silica network, thus making the processing of glass easy.

Conventionally, glasses are formed by melting an appropriate composition and then casting or drawing the melt into a desired form. It is interesting to compare the viscosity values of liquid metals with glasses. Molten metals have about the same viscosity as that of water (about 10^{-3} Pa·s) and transform to a crystalline solid state in a discontinuous manner when cooled. The viscosity of glasses, however, falls slowly and continuously with temperature. The shaping of glass is carried out in the viscosity range of $1\times10^3 \sim 1\times10^6$ Pa·s. Polymers are formed in the range $1\times10^3 \sim 1\times10^5$ Pa·s. Perhaps the most important characteristic of a viscoelastic material is that its rheological properties are dependent on time. This characteristic is manifested very markedly by amorphous or noncrystalline materials such as polymers.

A viscoelastic substance has a viscous and an elastic component. Figure 9.6(a) shows the stress-strain curve of an ideal elastic material. The load and unload curves are the same, and the energy lost as heat per cycle is zero in this case. In practice, there is always present an anelastic (i.e., a time-dependent) component, with the result that the unload curve does not in fact follow the load curve. Energy equal to the shaded area in Figure 9.6(b) is dissipated in each cycle. This phenomenon is exploited in damping out vibrations. Some polymers and soft metals (e.g., lead) have a high damping capacity. In springs and bells, a high damping capacity is undesirable. For such applications, one uses materials such as bronze, spring steel, etc., which have a low damping capacity.

Elastic deformation is instantaneous, which means that total deformation (or strain) occurs the instant the stress is applied or released (i.e., the strain is independent of time). In addition, upon release of the external stress, the deformation is totally recovered—the specimen assumes its original

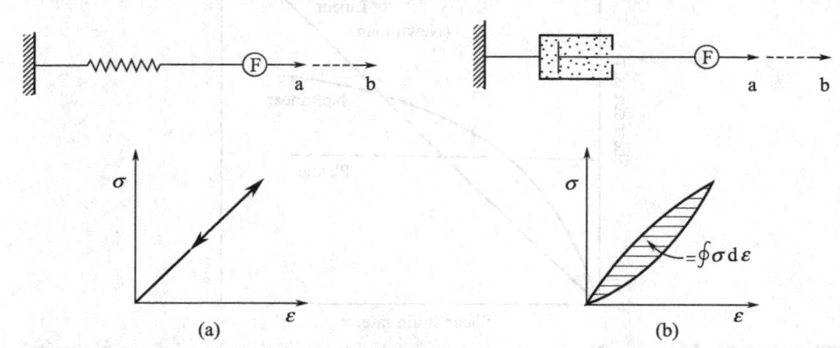

Figure 9.6 Stress-strain plots for (a) elastic behavior (no energy is lost during a load–unload cycle) and (b) viscoelastic behavior (energy equal to the shaded area is lost in a load–unload cycle).

dimensions. This behavior is represented in Figure 9.7(b) as strain versus time for the instantaneous load-time curve, shown in Figure 9.7(a).

By way of contrast, for totally viscous behavior, deformation or strain is not instantaneous; that is, in response to an applied stress, deformation is delayed or dependent on time. Also, this deformation is not reversible or completely recovered after the stress is released. This phenomenon is demonstrated in Figure 9.7(d).

For the intermediate viscoelastic behavior, the imposition of a stress in the manner of Figure 9.7a results in an instantaneous elastic strain, which is followed by a viscous, time-dependent strain, a form of anelasticity; this behavior is illustrated in Figure 9.7(c).

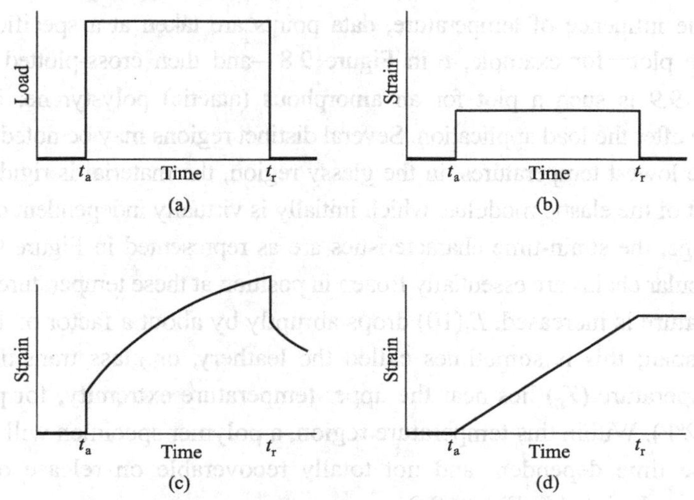

Figure 9.7 (a) Load versus time, where load is applied instantaneously at time t_a and released at t_r. For the load-time cycle in (a), the strain-versus-time responses are for totally elastic (b), viscoelastic (c), and viscous (d) behaviors.

A familiar example of these viscoelastic extremes is found in a silicone polymer that is sold as a novelty and known by some as "silly putty." When rolled into a ball and dropped onto a horizontal surface, it bounces elastically—the rate of deformation during the bounce is very rapid. On the other hand, if pulled in tension with a gradually increasing applied stress, the material elongates or flows like a highly viscous liquid. For this and other viscoelastic materials, the rate of strain determines whether the deformation is elastic or viscous.

Viscoelastic Relaxation Modulus

The viscoelastic behavior of polymeric materials is dependent on both time and temperature; several experimental techniques may be used to measure and quantify this behavior. Stress relaxation measurements represent one possibility. With these tests, a specimen is initially strained rapidly in tension to a predetermined and relatively low strain level. The stress necessary to maintain this strain is measured as a function of time, while temperature is held constant. Stress is found to decrease with time due to molecular relaxation processes that take place within the polymer. We may define a relaxation modulus $E_r(t)$, a time-dependent elastic modulus for viscoelastic polymers, as

$$E_r(t) = \frac{\sigma(t)}{\varepsilon_0} \tag{9.5}$$

Where $\sigma(t)$ is the measured time-dependent stress and ε_0 is the strain level, which is maintained constant.

Furthermore, the magnitude of the relaxation modulus is a function of temperature; and to more fully characterize the viscoelastic behavior of a polymer, isothermal stress relaxation measurements must be conducted over a range of temperatures. Figure 9.6 is a schematic log $E_r(t)$-versus-log time plot for a polymer that exhibits viscoelastic behavior. Curves generated at a variety of temperatures are included. Key features of this plot are that:

(1) the magnitude of $E_r(t)$decreases with time (corresponding to the decay of stress, Equation 9.1),

(2) the curves are displaced to lower $E_r(t)$levels with increasing temperature.

To represent the influence of temperature, data points are taken at a specific time from the log $E_r(t)$-versus-log time plot—for example, t_1 in Figure 9.8—and then cross-plotted as log $E_r(t)$ versus temperature. Figure 9.9 is such a plot for an amorphous (atactic) polystyrene; in this case, t_1 was arbitrarily taken 10 s after the load application. Several distinct regions may be noted on the curve shown in this figure. At the lowest temperatures, in the glassy region, the material is rigid and brittle, and the value of $E_r(10)$is that of the elastic modulus, which initially is virtually independent of temperature. Over this temperature range, the strain-time characteristics are as represented in Figure 9.9. On a molecular level, the long molecular chains are essentially frozen in position at these temperatures.

As the temperature is increased, $E_r(10)$ drops abruptly by about a factor of 1×10^3 within a 20℃ (35℉)temperature span; this is sometimes called the leathery, or glass transition region, and the glass transition temperature (T_g) lies near the upper temperature extremity; for polystyrene (Figure 9.9), T_g=100℃ (212℉). Within this temperature region, a polymer specimen will be leathery; that is, deformation will be time dependent and not totally recoverable on release of an applied load, characteristics that are depicted in Figure 9.9.

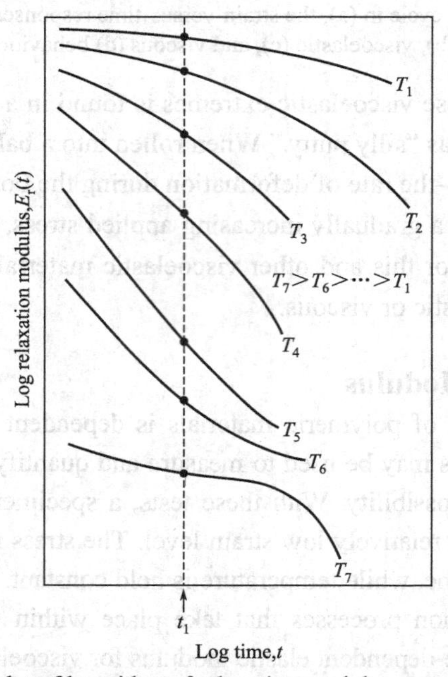

Figure 9.8 Schematic plot of logarithm of relaxation modulus versus logarithm of time for a viscoelastic polymer; isothermal curves are generated at temperatures T_1 through T_7. The temperature dependence of the relaxation modulus is represented as log $E_r(t_1)$ versus temperature.

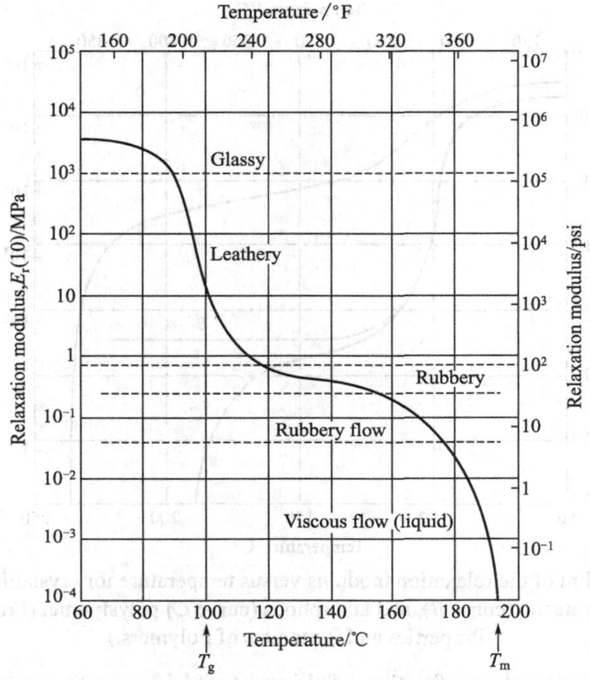

Figure 9.9 Logarithm of the relaxation modulus versus temperature for amorphous polystyrene, showing the five different regions of viscoelastic behavior. (From A. V. Tobolsky, Properties and Structures of Polymers.)

Within the rubbery plateau temperature region (Figure 9.9), the material deforms in a rubbery manner; here, both elastic and viscous components are present, and deformation is easy to produce because the relaxation modulus is relatively low. The final two high-temperature regions are rubbery flow and viscous flow. Upon heating through these temperatures, the material experiences a gradual transition to a soft rubbery state, and finally to a viscous liquid. In the rubbery flow region, the polymer is a very viscous liquid that exhibits both elastic and viscous flow components. Within the viscous flow region, the modulus decreases dramatically with increasing temperature; again, the strain–time behavior is as represented in Figure 9.9(d). From a molecular standpoint, chain motion intensifies so greatly that for viscous flow, the chain segments experience vibration and rotational motion largely independent of one another. At these temperatures, any deformation is entirely viscous and essentially no elastic behavior occurs.

Normally, the deformation behavior of a viscous polymer is specified in terms of viscosity, a measure of a material's resistance to flow by shear forces. The rate of stress application also influences the viscoelastic characteristics. Increasing the loading rate has the same influence as lowering temperature. The log $E_r(10)$-versus -temperature behavior for polystyrene materials having several molecular configurations is plotted in Figure 9.10. The curve for the amorphous material (curve *C*) is the same as in Figure 9.11. For a lightly crosslinked atactic polystyrene (curve *B*), the rubbery region forms a plateau that extends to the temperature at which the polymer decomposes; this material will not experience melting. For increased crosslinking, the magnitude of the plateau $E_r(10)$value will also increase. Rubber or elastomeric materials display this type of behavior and are ordinarily used at temperatures within this plateau range.

Also shown in Figure 9.10 is the temperature dependence for an almost totally crystalline isotactic polystyrene (curve *A*).The decrease in $E_r(10)$at T_g is much less pronounced than the other polystyrene

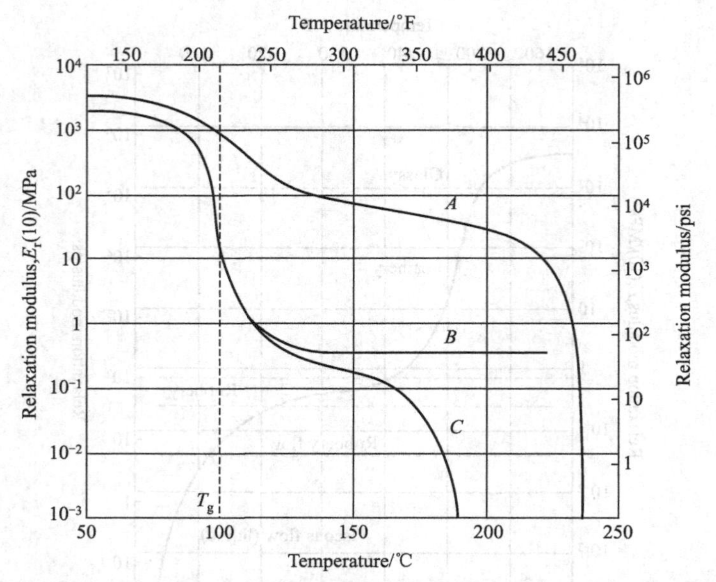

Figure 9.10 Logarithm of the relaxation modulus versus temperature for crystalline isotactic (curve *A*), lightly crosslinked atactic (curve *B*), and amorphous (curve *C*) polystyrene. (From A. V. Tobolsky, Properties and Structures of Polymers.)

materials since only a small volume fraction of this material is amorphous and experiences the glass transition. Furthermore, the relaxation modulus is maintained at a relatively high value with increasing temperature until its melting temperature T_m is approached. From Figure 9.10, the melting temperature of this isotactic polystyrene is about 240 ℃ (460℉).

Viscoelastic Creep

Many polymeric materials are susceptible to time-dependent deformation when the stress level is maintained constant; such deformation is termed viscoelastic creep. This type of deformation may be significant even at room temperature and under modest stresses that lie below the yield strength of the material. For example, automobile tires may develop flat spots on their contact surfaces when the automobile is parked for prolonged time periods. Creep tests on polymers are conducted in the same manner as for metals; that is, a stress (normally tensile) is applied instantaneously and is maintained at a constant level while strain is measured as a function of time. Furthermore, the tests are performed under isothermal conditions. Creep results are represented as a time-dependent creep modulus $E_c(t)$, defined by

$$E_c(t) = \frac{\sigma_0}{\varepsilon(t)} \tag{9.6}$$

Wherein σ_0 is the constant applied stress and is the time-dependent strain. The creep modulus is also temperature sensitive and diminishes with increasing temperature. With regard to the influence of molecular structure on the creep characteristics, as a general rule the susceptibility to creep decreases [i.e., $E_c(t)$ increases] as the degree of crystallinity increases.

9.5 Fracture Of Polymers

The fracture strengths of polymeric materials are low relative to those of metals and ceramics. As a general rule, the mode of fracture in thermosetting polymers (heavily crosslinked networks) is brittle. In simple terms, during the fracture process, cracks form at regions where there is a localized stress concentration (i.e., scratches, notches, and sharp flaws). As with metals, the stress is amplified at the tips of these cracks leading to crack propagation and fracture. Covalent bonds in the network or crosslinked structure are severed during fracture.

For thermoplastic polymers, both ductile and brittle modes are possible, and many of these materials are capable of experiencing a ductile-to-brittle transition. Factors that favor brittle fracture are a reduction in temperature, an increase in strain rate, the presence of a sharp notch, increased specimen thickness, and any modification of the polymer structure that raises the glass transition temperature T_g. Glassy thermoplastics are brittle below their glass transition temperatures.

However, as the temperature is raised, they become ductile in the vicinity of their T_g s and experience plastic yielding prior to fracture. This behavior is demonstrated by the stress–strain characteristics of poly (methyl methacrylate) in Figure 9.3. At 4°C, PMMA is totally brittle, whereas at 60°C it becomes extremely ductile.

One phenomenon that frequently precedes fracture in some thermoplastic polymers is *crazing*. Associated with crazes are regions of very localized plastic deformation, which lead to the formation of small and interconnected microvoids [Figure 9.11(a)].

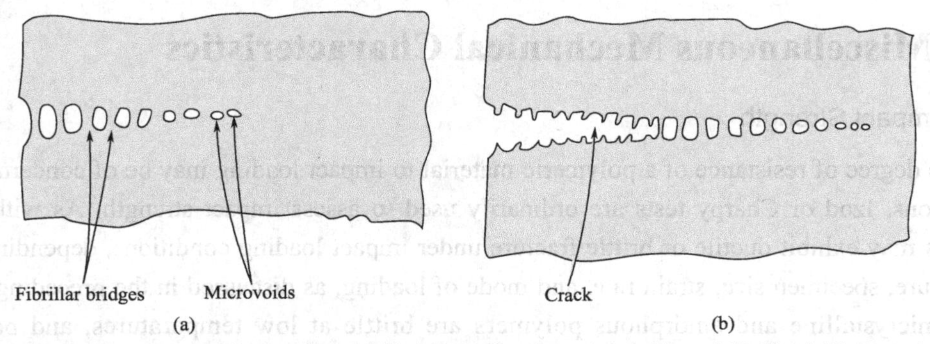

Figure 9.11 Schematic drawings of (a) a craze showing microvoids and fibrillar bridges, and (b) a craze followed by a crack. (From J.W. S. Hearle, Polymers and Their Properties, Vol. 1, Fundamentals of Structure and Mechanics, Ellis Horwood, Ltd., Chichester, West Sussex, England, 1982.)

Fibrillar bridges form between these microvoids wherein molecular chains become oriented as in Figure 9.15. If the applied tensile load is sufficient, these bridges elongate and break, causing the microvoids to grow and coalesce. As the microvoids coalesce, cracks begin to form, as demonstrated in Figure 9.11(b). A craze is different from a crack in that it can support a load across its face. Furthermore, this process of craze growth prior to cracking absorbs fracture energy and effectively increases the fracture toughness of the polymer. In glassy polymers, the cracks propagate with little craze formation resulting in low fracture toughnesses. Crazes form at highly stressed regions associated with scratches, flaws, and molecular inhomogeneities; in addition, they propagate perpendicular to the applied tensile stress, and typically are 5 μm or less thick. Figure 9.12 is a photomicrograph in which a craze is shown.

Principles of fracture mechanics developed in Section 9.5 also apply to brittle and quasi-brittle polymers; the susceptibility of these materials to fracture when a crack is present may be expressed in terms of the plane strain fracture toughness. The magnitude of K_{Ic} will depend on characteristics of the polymer (i.e., molecular weight, percent crystallinity, etc.) as well as temperature, strain rate, and the external environment.

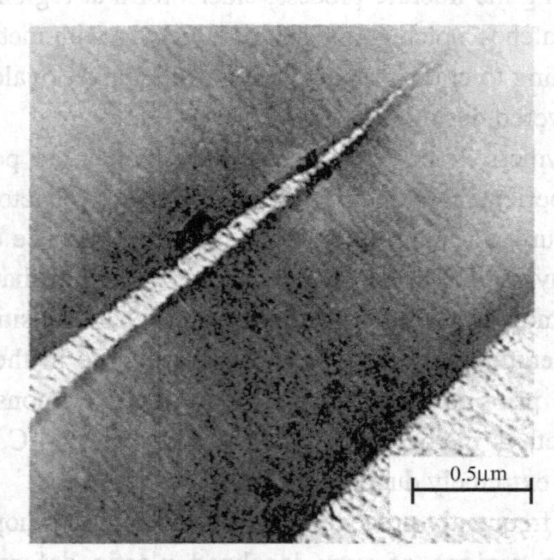

Figure 9.12 Photomicrograph of a craze in poly(phenylene oxide). (From R. P. Kambour and R. E. Robertson, " The Mechanical Properties of Plastics," in Polymer Science, A Materials Science Handbook, A. D. Jenkins, Editor.)

9.6 Miscellaneous Mechanical Characteristics

9.6.1 Impact Strength

The degree of resistance of a polymeric material to impact loading may be of concern in some applications. Izod or Charpy tests are ordinarily used to assess impact strength. As with metals, polymers may exhibit ductile or brittle fracture under impact loading conditions, depending on the temperature, specimen size, strain rate, and mode of loading, as discussed in the preceding section. Both semicrystalline and amorphous polymers are brittle at low temperatures, and both have relatively low impact strengths. However, they experience a ductile-to-brittle transition over a relatively narrow temperature range, similar to that shown for a steel in Figure 9.15. Of course, impact strength undergoes a gradual decrease at still higher temperatures as the polymer begins to soften. Ordinarily, the two impact characteristics most sought after are a high impact strength at the ambient temperature and a ductile-to-brittle transition temperature that lies below room temperature.

9.6.2 Fatigue

Polymers may experience fatigue failure under conditions of cyclic loading. As with metals, fatigue occurs at stress levels that are low relative to the yield strength. Fatigue testing in polymers has not been nearly as extensive as with metals; however, fatigue data are plotted in the same manner for both types of material, and the resulting curves have the same general shape. Fatigue curves for several common polymers are shown in Figure 9.13, as stress versus the number of cycles to failure (on a logarithmic scale). Some polymers have a fatigue limit (a stress level at which the stress at failure becomes independent of the number of cycles); others do not appear to have such a limit. As would be expected, fatigue strengths and fatigue limits for polymeric materials are much lower than for metals.

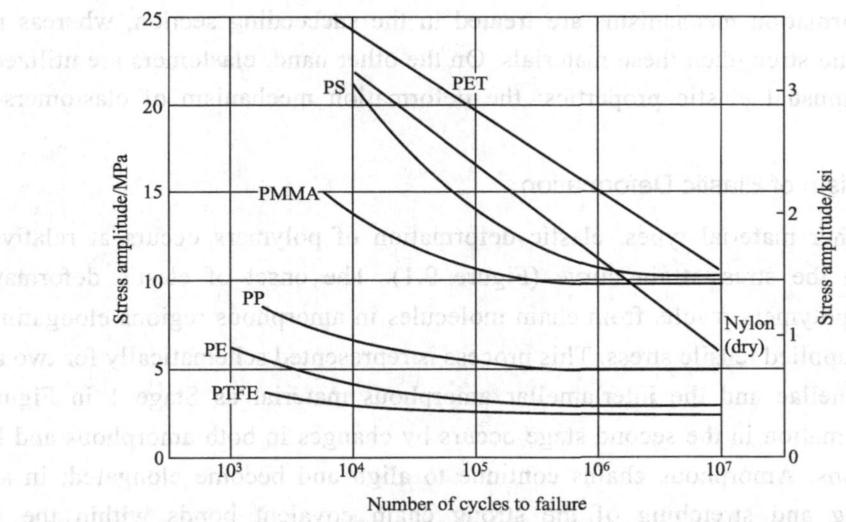

Figure 9.13 Fatigue curves (stress amplitude versus the number of cycles to failure) for poly(ethylene terephthalate) (PET), nylon, polystyrene (PS), poly(methyl methacrylate) (PMMA), polypropylene (PP), polyethylene (PE), and polytetrafluoroethylene (PTFE). The testing frequency was 30 Hz (From M. N. Riddell, "A Guide to Better Testing of Plastics," *Plast. Eng.*, Vol. 30, No. 4, p. 78, 1974).

The fatigue behavior of polymers is much more sensitive to loading frequency than for metals. Cycling polymers at high frequencies and/or relatively large stresses can cause localized heating; consequently, failure may be due to a softening of the material rather than as a result of typical fatigue processes.

9.6.3 Tear Strength and Hardness

Other mechanical properties that are sometimes influential in the suitability of a polymer for some particular application include tear resistance and hardness. The ability to resist tearing is an important property of some plastics, especially those used for thin films in packaging. Tear strength, the mechanical parameter that is measured, is the energy required to tear apart a cut specimen that has a standard geometry. The magnitude of tensile and tear strengths are related.

As with metals, hardness represents a material's resistance to scratching, penetration, marring, and so on. Polymers are softer than metals and ceramics, and most hardness tests are conducted by penetration techniques similar to those described for metals in Section 6.10. Rockwell tests are frequently used for polymers. Other indentation techniques employed are the Durometer and Barcol.

9.7 Mechanisms of Deformation and for Strengthening of Polymers

An understanding of deformation mechanisms of polymers is important in order for us to be able to manage the mechanical characteristics of these materials. In this regard, deformation models for two different types of polymers—semicrystalline and elastomeric—deserve our attention.The stiffness and strength of semicrystalline materials are often important considerations; elastic and plastic deformation mechanisms are treated in the succeeding section, whereas methods used to stiffen and strengthen these materials. On the other hand, elastomers are utilized on the basis of their unusual elastic properties; the deformation mechanism of elastomers is also treated.

9.7.1 Mechanism of Elastic Deformation

As with other material types, elastic deformation of polymers occurs at relatively low stress levels on the stress-strain curve (Figure 9.1). The onset of elastic deformation for semicrystalline polymers results from chain molecules in amorphous regions elongating in the direction of the applied tensile stress. This process is represented schematically for two adjacent chain-folded lamellae and the interlamellar amorphous material as Stage 1 in Figure 9.14. Continued deformation in the second stage occurs by changes in both amorphous and lamellar crystalline regions. Amorphous chains continue to align and become elongated; in addition, there is bending and stretching of the strong chain covalent bonds within the lamellar crystallites. This leads to a slight, reversible increase in the lamellar crystallite thickness as indicated by Δt in Figure 9.14(c).

In as much as semicrystalline polymers are composed of both crystalline and amorphous regions, they may, in a sense, be considered composite materials. As such, the elastic modulus may be taken as some combination of the moduli of crystalline and amorphous phases.

9.7.2 Mechanism of Plastic Deformation

The transition from elastic to plastic deformation occurs in Stage 3 of Figure 9.15 [Note that Figure 9.14(c) is identical to Figure 9.15(a)] During Stage 3, adjacent chains in the lamellae slide past one another [Figure 9.15(b)]; this results in tilting of the lamellae so that the chain folds become more aligned with the tensile axis. Any chain displacement is resisted by relatively weak secondary or van der Waals bonds. Crystalline block segments separate from the lamellae, in Stage 4 [Figure 9.15(c)], with the segments attached to one another by tie chains. In the final stage, Stage 5, the blocks and tie chains become oriented in the direction of the tensile axis [Figure 9.15(d)]. Thus, appreciable tensile deformation of semicrystalline polymers produces a highly oriented structure. This process of orientation is referred to as drawing, and is commonly used to improve the mechanical properties of polymer fibers and films.During deformation the spherulites experience shape changes for moderate levels of elongation.

However, for large deformations, the spherulitic structure is virtually destroyed. Also, to a degree, the processes represented in Figure 9.15 are reversible. That is, if deformation is terminated at some arbitrary stage, and the specimen is heated to an elevated temperature near its melting point (i.e., is annealed), the material will recrystallize to again form a spherulitic structure. Furthermore, the specimen will tend to shrink back, in part, to the dimensions it had prior to deformation. The extent of this shape and structural recovery will depend on the annealing temperature and also the degree of elongation.

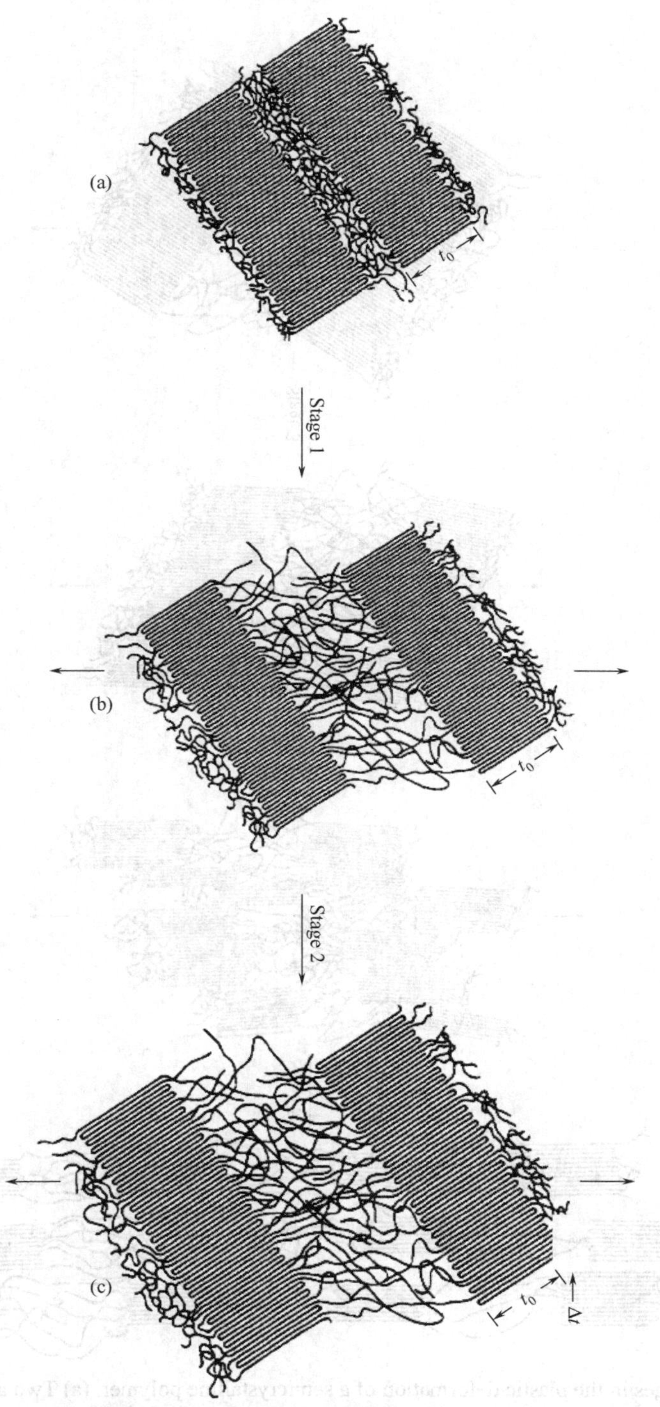

Figure 9.14 Stages in the elastic deformation of a semicrystalline polymer. (a) Two adjacent chain-folded lamellae and interlamellar amorphous material before deformation. (b) Elongation of amorphous tie chains during the first stage of deformation. (c) Increase in lamellar crystallite thickness (which is reversible) due to bending and stretching of chains in crystallite regions (From Schultz, Jerold M., *Polymer Materials Science,* 1st edition, © 1974, pp. 500, 501. Adapted by permission of Pearson Education, Inc., Upper Saddle River, NJ) .

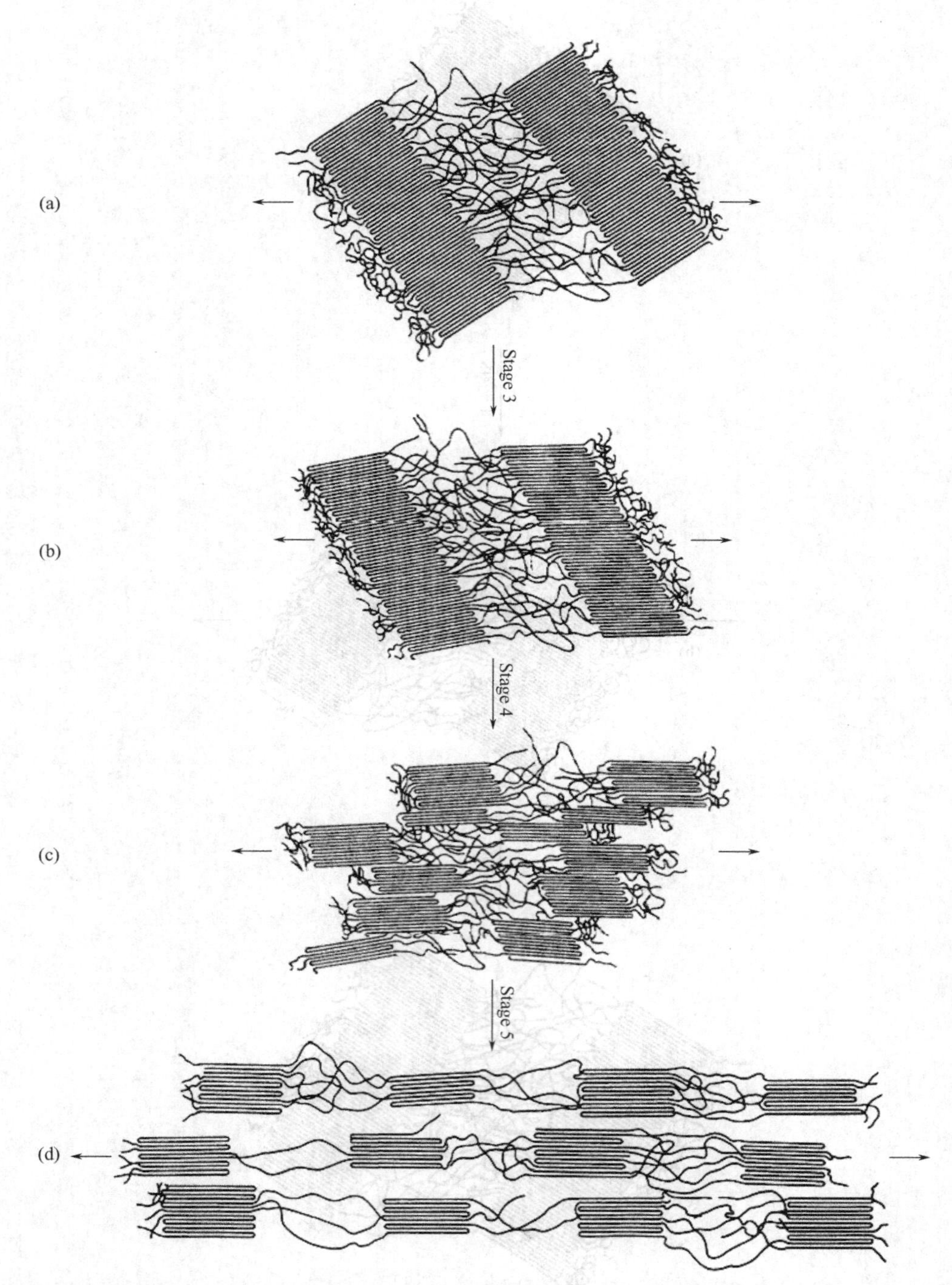

Figure 9.15 Stages in the plastic deformation of a semicrystalline polymer. (a) Two adjacent chain-folded lamellae and interlamellar amorphous material after elastic deformation [also shown as Figure 9.15(c)]. (b) Tilting of lamellar chain folds. (c) Separation of crystalline block segments. (d) Orientation of block segments and tie chains with the tensile axis in the final plastic deformation stage (From Schultz, Jerold M., Polymer Materials Science, 1st edition, © 1974, pp. 500, 501. Adapted by permission of Pearson Education, Inc., Upper Saddle River, NJ) .

9.8 Factors That Influence The Mechanical Properties Of Semicrystalline

9.8.1 Polymers

A number of factors influence the mechanical characteristics of polymeric materials. For example, we have already discussed the effects of temperature and strain rate on stress–strain behavior (Section 9.2, Figure 9.3). Again, increasing the temperature or diminishing the strain rate leads to a decrease in the tensile modulus, a reduction in tensile strength, and an enhancement of ductility.

In addition, several structural/processing factors have decided influences on the mechanical behavior (i.e., strength and modulus) of polymeric materials. An increase in strength results whenever any restraint is imposed on the process illustrated in Figure 9.15; for example, extensive chain entanglements or a significant degree of intermolecular bonding inhibit relative chain motions. It should be noted that even though secondary intermolecular (e.g., van der Waals) bonds are much weaker than the primary covalent ones, significant intermolecular forces result from the formation of large numbers of van der Waals interchain bonds. Furthermore, the modulus rises as both the secondary bond strength and chain alignment increase.

As a result, polymers with polar groups will have stronger secondary bonds and a larger elastic modulus. We now discuss how several structural/processing factors [viz. molecular weight, degree of crystallinity, predeformation (drawing), and heat treating] affect the mechanical behavior of polymers.

9.8.2 Molecular Weight

The magnitude of the tensile modulus does not seem to be directly influenced by molecular weight. On the other hand, for many polymers it has been observed that tensile strength increases with increasing molecular weight. Mathematically, *TS* is a function of the number-average molecular weight according to

$$TS = TS_{\infty} - \frac{A}{M_n} \quad (9.7)$$

where TS_{∞} is the tensile strength at infinite molecular weight and A is a constant. The behavior described by this equation is explained by increased chain entanglements with rising M_n.

9.8.3 Degree of Crystallinity

For a specific polymer, the degree of crystallinity can have a rather significant influence on the mechanical properties, since it affects the extent of the intermolecular secondary bonding. For crystalline regions in which molecular chains are closely packed in an ordered and parallel arrangement, extensive secondary bonding ordinarily exists between adjacent chain segments. This secondary bonding is much less prevalent in amorphous regions, by virtue of the chain misalignment. As a consequence, for semicrystalline polymers, tensile modulus increases significantly with degree of crystallinity. For example, for polyethylene, the modulus increases approximately an order of magnitude as the crystallinity fraction is raised from 0.3 to 0.6.

Furthermore, increasing the crystallinity of a polymer generally enhances its strength; in addition, the material tends to become more brittle. The effects of both percent crystallinity and molecular weight on the physical state of polyethylene are represented in Figure 9.16.

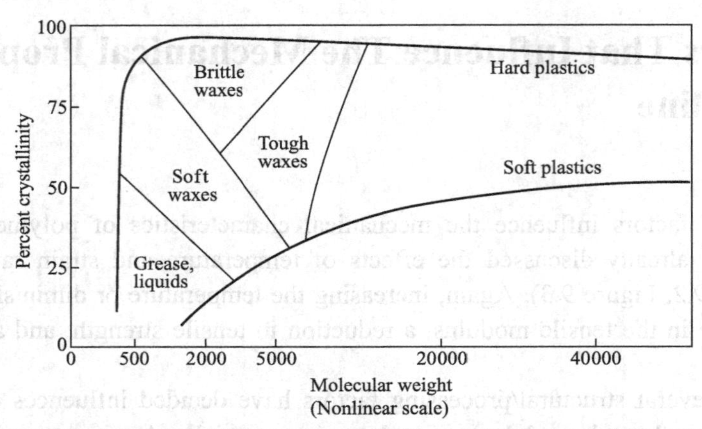

Figure 9.16 The influence of degree of crystallinity and molecular weight on the physical characteristics of polyethylene (From R. B. Richards, "Polyethylene—Structure, Crystallinity and Properties," J. Appl. hem., 1, 370, 1951) .

9.8.4 Predeformation by Drawing

On a commercial basis, one of the most important techniques used to improve mechanical strength and tensile modulus is to permanently deform the polymer in tension. This procedure is sometimes termed drawing, and corresponds to the neck extension process illustrated schematically in Figure 9.10. In terms of property alterations, drawing is the polymer analog of strain hardening in metals. It is an important stiffening and strengthening technique that is employed in the production of fibers and films. During drawing the molecular chains slip past one another and become highly oriented; for semicrystalline materials the chains assume conformations similar to that represented schematically in Figure 9.15(d).

Degrees of strengthening and stiffening will depend on the extent of deformation (or extension) of the material. Furthermore, the properties of drawn polymers are highly anisotropic. For those materials drawn in uniaxial tension, tensile modulus and strength values are significantly greater in the direction of deformation than in other directions. Tensile modulus in the direction of drawing may be enhanced by up to approximately a factor of three relative to the undrawn material. At an angle of 45° from the tensile axis the modulus is a minimum; at this orientation the modulus has a value on the order of one-fifth that of the undrawn polymer.

Tensile strength parallel to the direction of orientation may be improved by a factor of at least two to five relative to that of the unoriented material. On the other hand, perpendicular to the alignment direction, tensile strength is reduced by on the order of one-third to one-half.

For an amorphous polymer that is drawn at an elevated temperature, the oriented molecular structure is retained only when the material is quickly cooled to the ambient; this procedure gives rise to the strengthening and stiffening effects described in the previous paragraph. On the other hand, if, after stretching, the polymer is held at the temperature of drawing, molecular chains relax and assume random conformations characteristic of the predeformed state; as a consequence, drawing will have no effect on the mechanical characteristics of the material.

9.8.5 Heat Treating

Heat treating (or annealing) of semicrystalline polymers can lead to an increase in the percent crystallinity, and crystallite size and perfection, as well as modifications of the spherulite structure.

For undrawn materials that are subjected to constant-time heat treatments, increasing the annealing temperature leads to the following:

(1) an increase in tensile modulus,

(2) an increase in yield strength,

(3) a reduction in ductility. Note that these annealing effects are opposite to those typically observed for metallic materials—i.e., weakening, softening, and enhanced ductility.

For some polymer fibers that have been drawn, the influence of annealing on the tensile modulus is contrary to that for undrawn materials—that is, modulus decreases with increased annealing temperature due to a loss of chain orientation and strain-induced crystallinity.

9.9 Deformation Of Elastomers

One of the fascinating properties of the elastomeric materials is their rubber-like elasticity. That is, they have the ability to be deformed to quite large deformations, and then elastically spring back to their original form. This results from crosslinks in the polymer that provide a force to restore the chains to their undeformed conformations. Elastomeric behavior was probably first observed in natural rubber; however, the past few years have brought about the synthesis of a large number of elastomers with a wide variety of properties. Typical stress-strain characteristics of elastomeric materials are displayed in Figure 9.1, curve *C*. Their moduli of elasticity are quite small and, furthermore, vary with strain since the stress–strain curve is nonlinear.In an unstressed state, an elastomer will be amorphous and composed of crosslinked molecular chains that are highly twisted, kinked, and coiled. Elastic deformation, upon application of a tensile load, is simply the partial uncoiling, untwisting, and straightening, and the resultant elongation of the chains in the stress direction, a phenomenon represented in Figure 9.17. Upon release of the stress, the chains spring back to their prestressed conformations, and the macroscopic piece returns to its original shape.

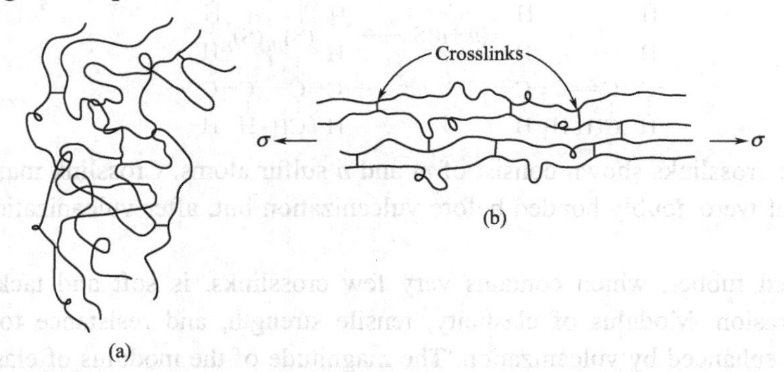

Figure 9.17 Schematic representation of crosslinked polymer chain molecules (a) in an unstressed state and (b) during elastic deformation in response to an applied tensile stress (Adapted from Z. D. Jastrzebski, The Nature and Properties of Engineering Materials, 3rd edition).

Part of the driving force for elastic deformation is a thermodynamic parameter called *entropy,* which is a measure of the degree of disorder within a system; entropy increases with increasing disorder. As an elastomer is stretched and the chains straighten and become more aligned, the system becomes more ordered. From this state, the entropy increases if the chains return to their original kinked and coiled contours. Two intriguing phenomena result from this entropic effect. First, when stretched, an elastomer experiences a rise in temperature; second, the modulus of elasticity increases with increasing temperature, which is opposite to the behavior found in other materials.

Several criteria must be met for a polymer to be elastomeric:

(1) It must not easily crystallize; elastomeric materials are amorphous, having molecular chains that are naturally coiled and kinked in the unstressed state.

(2) Chain bond rotations must be relatively free for the coiled chains to readily respond to an applied force.

(3) For elastomers to experience relatively large elastic deformations, the onset of plastic deformation must be delayed. Restricting the motions of chains past one another by crosslinking accomplishes this objective.The crosslinks act as anchor points between the chains and prevent chain slippage from occurring; the role of crosslinks in the deformation process is illustrated in Figure 9.17. Crosslinking in many elastomers is carried out in a process called vulcanization, to be discussed below.

(4) Finally, the elastomer must be above its glass transition temperature. The lowest temperature at which rubber-like behavior persists for many of the common elastomers is between −50 °C and −90 °C (−60 °F and −130 °F).Below its glass transition temperature, an elastomer becomes brittle such that its stress–strain behavior resembles curve *A* in Figure 9.1.

Vulcanization

The crosslinking process in elastomers is called vulcanization, which is achieved by a nonreversible chemical reaction, ordinarily carried out at an elevated temperature. In most vulcanizing reactions, sulfur compounds are added to the heated elastomer; chains of sulfur atoms bond with adjacent polymer backbone chains and crosslink them, which is accomplished according to the following reaction:

$$\begin{array}{c} \begin{array}{cccc} \mathrm{H} & \mathrm{CH_3} & \mathrm{H} & \mathrm{H} \\ | & | & | & | \\ -\mathrm{C}- & \mathrm{C}= & \mathrm{C}- & \mathrm{C}- \\ | & & & | \\ \mathrm{H} & & & \mathrm{H} \\ \\ \mathrm{H} & & & \mathrm{H} \\ | & & & | \\ -\mathrm{C}- & \mathrm{C}= & \mathrm{C}- & \mathrm{C}- \\ | & | & | & | \\ \mathrm{H} & \mathrm{CH_3} & \mathrm{H} & \mathrm{H} \end{array} + (m+n)\mathrm{S} \longrightarrow \begin{array}{cccc} \mathrm{H} & \mathrm{CH_3} & \mathrm{H} & \mathrm{H} \\ | & | & | & | \\ -\mathrm{C}- & \mathrm{C}- & \mathrm{C}- & \mathrm{C}- \\ | & | & | & | \\ \mathrm{H} & (\mathrm{S})_m & (\mathrm{S})_n & \mathrm{H} \\ \mathrm{H} & | & | & \mathrm{H} \\ | & | & | & | \\ -\mathrm{C}- & \mathrm{C}- & \mathrm{C}- & \mathrm{C}- \\ | & | & | & | \\ \mathrm{H} & \mathrm{CH_3} & \mathrm{H} & \mathrm{H} \end{array} \end{array} \tag{9.8}$$

in which the two crosslinks shown consist of *m* and *n* sulfur atoms. Crosslink main chain sites are carbon atoms that were doubly bonded before vulcanization but, after vulcanization, have become singly bonded.

Unvulcanized rubber, which contains very few crosslinks, is soft and tacky and has poor resistance to abrasion. Modulus of elasticity, tensile strength, and resistance to degradation by oxidation are all enhanced by vulcanization. The magnitude of the modulus of elasticity is directly proportional to the density of the crosslinks.

Stress–strain curves for vulcanized and unvulcanized natural rubber are presented in Figure 9.18. To produce a rubber that is capable of large extensions without rupture of the primary chain bonds, there must be relatively few crosslinks, and these must be widely separated. Useful rubbers result when about 1 to 5 parts (by weight) of sulfur are added to 100 parts of rubber. This corresponds to about one crosslink for every 10 to 20 repeat units. Increasing the sulfur content further hardens the rubber and also reduces its extensibility. Also, since they are crosslinked, elastomeric materials are thermosetting in nature.

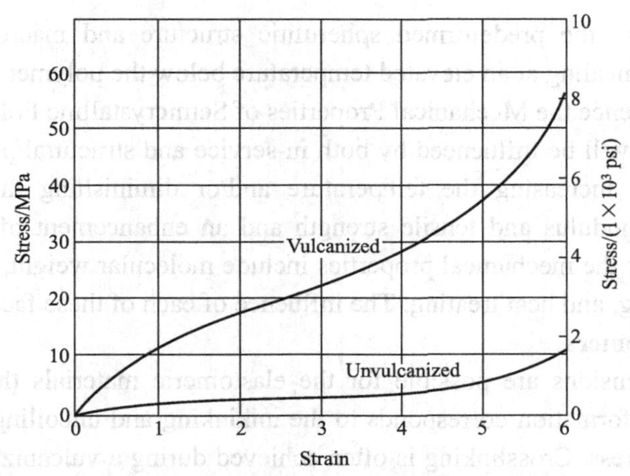

Figure 9.18 Stress–strain curves to 600% elongation for unvulcanized and vulcanized natural rubber

SUMMARY

Stress-Strain Behavior

On the basis of stress–strain behavior, polymers fall within three general classifications: brittle, plastic, and highly elastic. These materials are neither as strong nor as stiff as metals, and their mechanical properties are sensitive to changes in temperature and strain rate. However, their high flexibilities, low densities, and resistance to corrosion make them the materials of choice for many applications.

Viscoelastic Deformation

Viscoelastic mechanical behavior, being intermediate between totally elastic and totally viscous, is displayed by a number of polymeric materials. It is characterized by the relaxation modulus, a time-dependent modulus of elasticity. The magnitude of the relaxation modulus is very sensitive to temperature; critical to the in-service temperature range for elastomers is this temperature dependence.

Fracture of Polymers

Fracture strengths of polymeric materials are low relative to metals and ceramics. Both brittle and ductile fracture modes are possible, and some thermoplastic materials experience a ductile-to-brittle transition with a lowering of temperature, an increase in strain rate, and/or an alteration of specimen thickness or geometry. In some thermoplastics, the crack-formation process may be preceded by crazing; crazing can lead to an increase in ductility and toughness of the material.

Deformation of Semicrystalline Polymers

During the elastic deformation of a semicrystalline polymer having a spherulitic structure that is stressed in tension, the molecules in amorphous regions elongate in the stress direction. In addition, molecules in crystallites experience bending and stretching, which causes a slight increase in lamellar thickness.

The mechanism of plastic deformation for spherulitic polymers was also presented. Tensile deformation occurs in several stages as both amorphous tie chains and chain-folded block segments (which separate from the ribbon-like lamellae) become oriented with the tensile axis. Also, during deformation the shapes of spherulites are altered (for moderate deformations); relatively large degrees of deformation lead to a complete destruction of the spherulites to form highly aligned structures. Furthermore, the predeformed spherulitic structure and macroscopic shape may be partially restored by annealing at an elevated temperature below the polymer's melting temperature.

Factors That Influence the Mechanical Properties of Semicrystalline Polymers. The mechanical behavior of a polymer will be influenced by both in-service and structural/processing factors. With regard to the former, increasing the temperature and/or diminishing the strain rate leads to reductions in tensile modulus and tensile strength and an enhancement of ductility. In addition, other factors that affect the mechanical properties include molecular weight, degree of crystallinity, predeformation drawing, and heat treating. The influence of each of these factors was discussed.

Deformation of Elastomers

Large elastic extensions are possible for the elastomeric materials that are amorphous and lightly crosslinked. Deformation corresponds to the unkinking and uncoiling of chains in response to an applied tensile stress. Crosslinking is often achieved during a vulcanization process. Many of the elastomers are copolymers, whereas the silicone elastomers are really inorganic materials.

IMPORTANT TERMS AND CONCEPTS

Addition polymerization	Adhesive	Colorant
Condensation polymerization	Glass transition temperature	Ultrahigh molecular weight
Fiber	Filler	Flame retardant
Foam	Drawing	Liquid crystal polymer
Melting temperature	Molding	Plasticizer
Plastic	Relaxation modulus	Spinning
Stabilizer	Thermoplastic elastomer	Elastomer
polyethylene	Viscoelasticity	

REFERENCES

1. Billmeyer, F.W., Jr., Textbook of Polymer Science, 3rd edition, Wiley-Interscience, New York, 1984
2. Carraher, C. E., Jr., Seymour/Carraher's Polymer Chemistry, 6th edition, Marcel Dekker, New York, 2003
3. Harper, C. A. (Editor), Handbook of Plastics, Elastomers and Composites, 3rd edition, McGraw-Hill Professional Book Group, New York, 1996
4. Landel, R. F. (Editor), Mechanical Properties of Polymers and Composites, 2nd edition, Marcel Dekker, New York, 1994
5. McCrum, N. G., C. P. Buckley, and C. B. Bucknall, Principles of Polymer Engineering, 2nd edition, Oxford University Press, Oxford, 1997
6. Muccio, E. A., Plastic Part Technology, ASM International, Materials Park, OH, 1991
7. Powell, P.C., and A. J. Housz, Engineering with Polymers, 2nd edition, Nelson Thornes, Cheltenham, UK, 1998
8. Rosen, S. L., Fundamental Principles of Polymeric Materials, 2nd edition,Wiley, New York, 1993
9. Rudin, A., The Elements of Polymer Science and Engineering, 2nd edition, Academic Press, San Diego, 1998
10. Strong, A.B., Plastics: Materials and Processing, 3rd edition, Prentice Hall PTR, Paramus, IL, 2006
11. Tobolsky, A. V., Properties and Structures of Polymers, Wiley, New York, 1960. Advanced treatment. Ward, I. M. and J. Sweeney, An Introduction to the Mechanical Properties of Solid Polymers,2nd edition, John Wiley & Sons, Hoboken, NJ, 2004
12. Muccio, E.A., Plastics Processing Technology,ASM International, Materials Park, OH, 1994

QUESTIONS AND PROBLEMS

9.1 For thermoplastic polymers, cite five factors that favor brittle fracture. In your own words, briefly describe the phenomenon of viscoelasticity.

9.2 Briefly explain how each of the following influences the tensile modulus of a semicrystalline polymer and why:

(a) Molecular weight;

(b) Degree of crystallinity;

(c) Deformation by drawing;

(d) Annealing of an undeformed material;

(e) Annealing of a drawn material.

9.3 For each of the following pairs of polymers, do the following:

(1) state whether or not it is possible to decide whether one polymer has a higher tensile modulus than the other;

(2) if this is possible, note which has the higher tensile modulus and then cite the reason(s) for your choice;

(3) if it is not possible to decide, then state why.

(a) Branched and atactic poly(vinyl chloride) with a weight-average molecular weight of 100000 g/mol; linear and isotactic poly(vinyl chloride) having a weight-average molecular weight of 75000 g/mol.

(b) Random styrene-butadiene copolymer with 5% of possible sites crosslinked; block styrene-butadiene copolymer with 10% of possible sites crosslinked.

(c) Branched polyethylene with a numberaverage molecular weight of 100000 g/mol;atactic polypropylene with a number-average molecular weight of 150000 g/mol.

9.4 Would you expect the tensile strength of polychlorotrifluoroethylene to be greater than, the same as, or less than that of a polytetrafluoroethylene specimen having the same molecular weight and degree of crystallinity? Why?

9.5 Name the following polymer(s) that would be suitable for the fabrication of cups to contain hot coffee: polyethylene, polypropylene, poly(vinyl chloride), PET polyester, and polycarbonate.Why?

9.6 Make a schematic plot showing how the modulus of elasticity of an amorphous polymer depends on the glass transition temperature. Assume that molecular weight is held constant.

9.7 Cite the primary differences between addition and condensation polymerization techniques.

Chapter 9

IMPORTANT TERMS AND CONCEPTS

High Polymer	高分子聚合物	Stress Relaxation	应力松弛
Elastomer	高弹性塑料/合成橡胶/高弹体	Relaxation Modulus	松弛模量
Elastomer Plastic	弹性体塑料	Thermoplastic Elastomer	热塑性弹性材料
Condensation Polymerization	缩聚作用	Viscoelasticity	黏弹性
Fiber	纤维	Flame Retardant	耐燃剂
Foam	泡沫	Plasticizer	增塑剂/软化剂/增韧剂/塑化剂
Melting Temperature	熔化温度	Spinning	纺丝
Addition Polymerization	加聚作用	Thixotropic Agent	触变剂
Stabilizer	稳定剂	Inhomogeneity	非均一性/异质性
Polyethylene	聚乙烯	Homogeneity	均一性/同质性
Adhesive	黏合剂	Spherulite	球粒
Glass Transition Temperature	玻璃态转变温度	Heat Treatment	热处理
		Vulcanization	硫化/硬化作用
Filler	填料		
Drawing	拉拔		

Chapter 10

Mechanical properties of Composite Materials

Learning Objectives:

After careful study of this chapter you should be able to do the following:

1. Make schematic plots of the three characteristic stress–strain behaviors observed for composite materials.
2. Calculate longitudinal modulus and longitudinal strength for composites.
3. Understanding the failure modes in composites

10.1 Introduction

Many of our modern technologies require materials with unusual combinations of properties that cannot be met by the conventional metal alloys, ceramics, and polymeric materials. This is especially true for materials that are needed for aerospace, underwater, and transportation applications. For example, aircraft engineers are increasingly searching for structural materials that have low densities, are strong, stiff, and abrasion and impact resistant, and are not easily corroded. This is a rather formidable combination of characteristics. Frequently, strong materials are relatively dense; also, increasing the strength or stiffness generally results in a decrease in impact strength.

Material property combinations and ranges have been, and are yet being, extended by the development of composite materials. Generally speaking, a composite is considered to be any multiphase material that exhibits a significant proportion of the properties of both constituent phases such that a better combination of properties is realized. According to this principle of combined action, better property combinations are fashioned by the judicious combination of two or more distinct materials. Property trade-offs are also made for many composites.

Composites of sorts have already been discussed; these include multiphase metal alloys, ceramics, and polymers. For example, pearlitic steels have a microstructure consisting of alternating layers of ferrite and cementite. The ferrite phase is soft and ductile, whereas cementite is hard and very brittle. The combined mechanical characteristics of the pearlite (reasonably high ductility and strength) are superior to those of either of the constituent phases.

There are also a number of composites that occur in nature. For example, wood consists of strong and flexible cellulose fibers surrounded and held together by a stiffer material called lignin. Also, bone is a composite of the strong yet soft protein collagen and the hard, brittle mineral apatite.

A composite, in the present context, is a multiphase material that is artificially made, as opposed to one that occurs or forms naturally. In addition, the constituent phases must be chemically dissimilar and separated by a distinct interface. Thus, most metallic alloys and many ceramics do not fit this definition because their multiple phases are formed as a consequence of natural phenomena.

In designing composite materials, scientists and engineers have ingeniously combined various metals, ceramics, and polymers to produce a new generation of extraordinary materials. Most composites have been created to improve combinations of mechanical characteristics such as stiffness, toughness, and ambient and high-temperature strength.

Many composite materials are composed of just two phases; one is termed the matrix, which is continuous and surrounds the other phase, often called the dispersed phase. The properties of composites are a function of the properties of the constituent phases, their relative amounts, and the geometry of the dispersed phase."Dispersed phase geometry" in this context means the shape of the particles and the particle size, distribution, and orientation; these characteristics are represented in Figure 10.1.

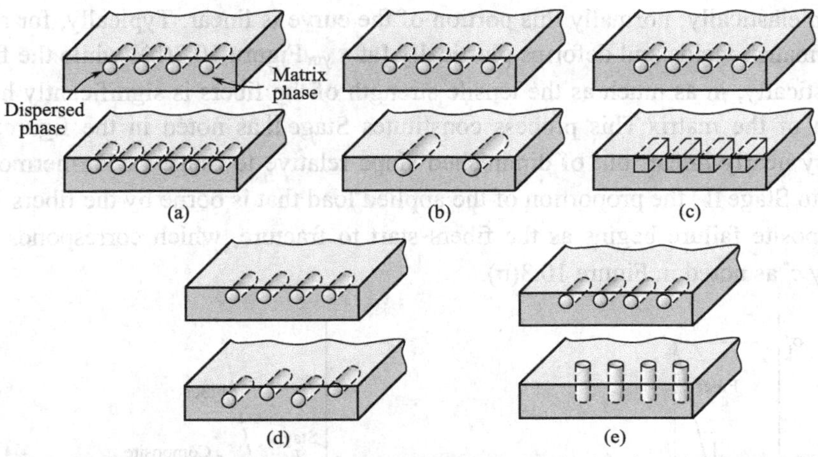

Figure 10.1 Schematic representations of the various geometrical and spatial characteristics of particles of the dispersed phase that may influence the properties of composites: (a) concentration, (b) size, (c) shape, (d) distribution, and (e) orientation.(From Richard A. Flinn and Paul K. Trojan, Engineering Materials and Their Applications, 4th edition.)

We may classify composites on the basis of the type of matrix employed in them—for example, polymer matrix composites (PMCs), metal matrix composites (MMCs), and ceramic matrix composites (CMCs). We may also classify composites on the basis of the type of reinforcement they employ (Figure 10.2):

(1) Particle reinforced composites.

(2) Short fiber, or whisker reinforced, composites.

(3) Continuous fiber, or sheet reinforced, MMCs.

(4) Laminate composite.

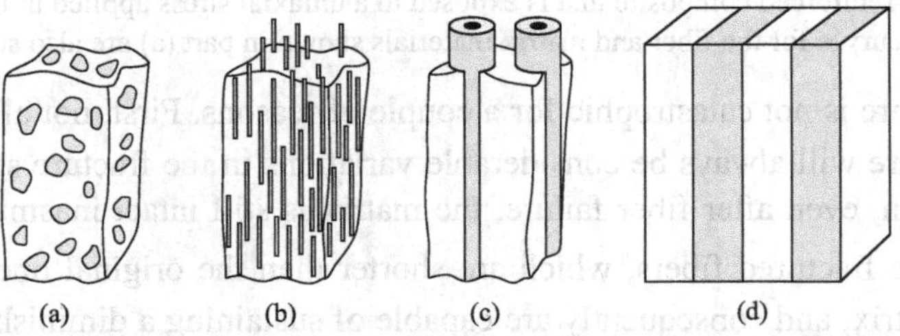

Figure 10.2 Different types of reinforcement for composites: (a) particle reinforcement; (b) short fiber reinforcement; (c) continuous fiber reinforcement; (d) laminate reinforcement.

10.2 Tensile Stress-Strain Behavior

To begin, assume the stress versus strain behaviors for fiber and matrix phases that are represented schematically in Figure 10.3(a); in this treatment we consider the fiber to be totally brittle and the matrix phase to be reasonably ductile. Also indicated in this figure are fracture strengths in tension for fiber and matrix, σ_f^* and σ_m^* respectively, and their corresponding fracture strains, ε_f^* and ε_m^* furthermore, it is assumed that $\varepsilon_f^* > \varepsilon_m^*$, which is normally the case.

A fiber-reinforced composite consisting of these fiber and matrix materials will exhibit the uniaxial stress–strain response illustrated in Figure 10.3(b); the fiber and matrix behaviors from Figure 10.3(a) are included to provide perspective. In the initial Stage I region, both fibers and matrix deform elastically; normally this portion of the curve is linear. Typically, for a composite of this type, the matrix yields and deforms plastically [at ε_{ym} Figure 10.3(b)] while the fibers continue to stretch elastically, in as much as the tensile strength of the fibers is significantly higher than the yield strength of the matrix.This process constitutes Stage II as noted in the figure; this stage is ordinarily very nearly linear, but of diminished slope relative to Stage I. Furthermore, in passing from Stage I to Stage II, the proportion of the applied load that is borne by the fibers increases. The onset of composite failure begins as the fibers start to fracture, which corresponds to a strain of approximately ε_f^* as noted in Figure 10.3(b).

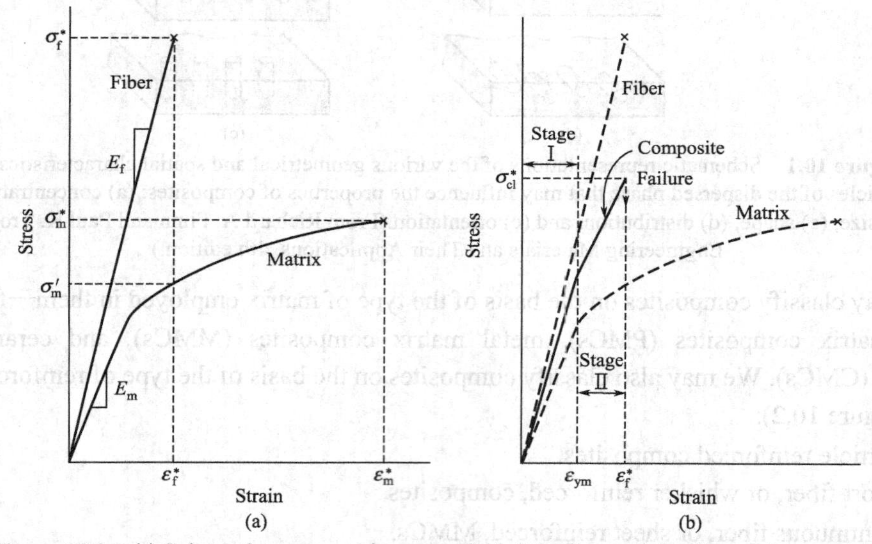

Figure 10.3 (a) Schematic stress–strain curves for brittle fiber and ductile matrix materials. Fracture stresses and strains for both materials are noted. (b) Schematic stress–strain curve for an aligned fiber-reinforced composite that is exposed to a uniaxial stress applied in the direction of alignment; curves for the fiber and matrix materials shown in part (a) are also superimposed.

Composite failure is not catastrophic for a couple of reasons. First, not all fibers fracture at the same time, since there will always be considerable variations in the fracture strength of brittle fiber materials. In addition, even after fiber failure, the matrix is still intact inasmuch as $\varepsilon_f^* < \varepsilon_m^*$ [Figure 10.3(a)]. Thus, these fractured fibers, which are shorter than the original ones, are still embedded within the intact matrix, and consequently are capable of sustaining a diminished load as the matrix continues to plastically deform.

Some critical fiber length is necessary for effective strengthening and stiffening of the composite material. This critical length is dependent on the fiber diameter d and its ultimate (or tensile) strength σ_f^*, and on the fiber–matrix bond strength (or the shear yield strength of the matrix, which ever is smaller) τ_c according to

$$l_c = \frac{\sigma_f^* d}{2\tau_c} \tag{10.1}$$

For a number of glass and carbon fiber–matrix combinations, this critical length is on the order of 1mm, which ranges between 20 and 150 times the fiber diameter.

When a stress equal to σ_f^* is applied to a fiber having just this critical length, the stress–position profile shown in Figure 10.4(a) results; that is, the maximum fiber load is achieved only at the axial center of the fiber. As fiber length l increases, the fiber reinforcement becomes more effective; this is demonstrated in Figure 10.4(b), a stress–axial position profile for $l > l_c$ when the applied stress is equal to the fiber strength. Figure 10.4(c) shows the stress–position profile for $l < l_c$.

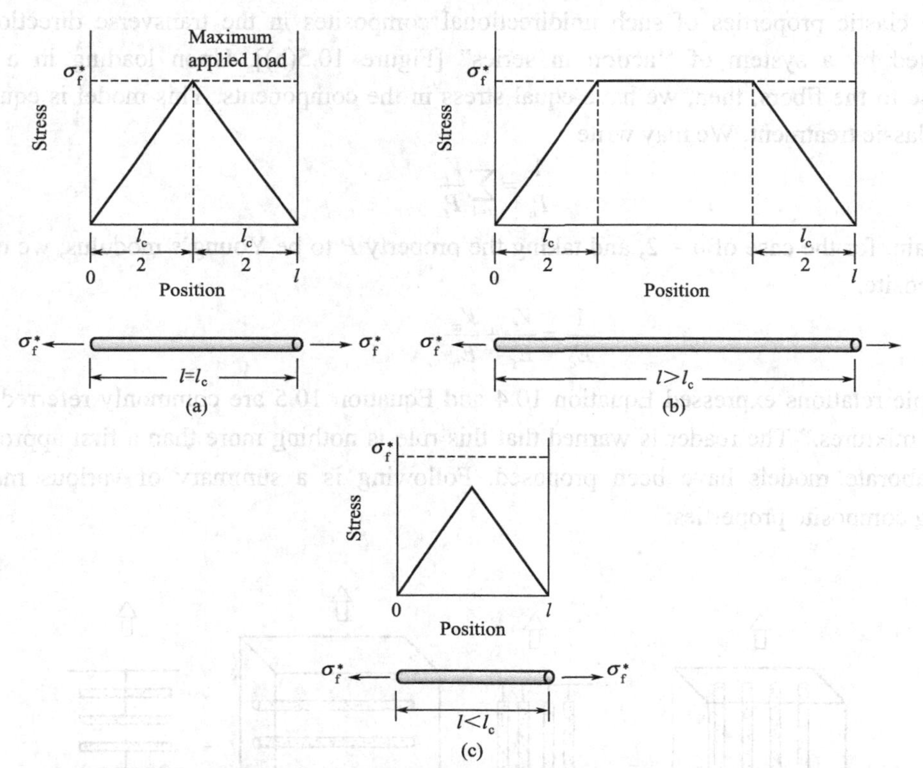

Figure 10.4 Stress–position profiles when fiber length l (a) is equal to the critical length (b) is greater than the critical length, and (c) is less than the critical length for a fiber–reinforced composite that is subjected to a tensile stress equal to the fiber tensile strength σ_f^*.

Fibers for which $l >> l_c$ (normally $l > 15l_c$) are termed continuous; discontinuous or short fibers have lengths shorter than this. For discontinuous fibers of lengths significantly less than l_c, the matrix deforms around the fiber such that there is virtually no stress transference and little reinforcement by the fiber. These are essentially the particulate composites as described above. To affect a significant improvement in strength of the composite, the fibers must be continuous.

10.2.1 Elastic Moduli

The simplest model for predicting the elastic properties of a fiber/reinforced composite is shown in Figure 10.5. In the longitudinal direction, the composite is represented by a system of "action in parallel" [Figure 10.5(a)]. For a load applied in the direction of the fibers, assuming equal deformation in the components, the two (or more) phases are viewed as being deformed in parallel. This is the classic case of Voigt's average, in which one has

$$P_c = \sum_{i=1}^{n} P_i V_i \tag{10.2}$$

where P is a property, V denotes volume fraction, and the subscripts c and i indicate, respectively, the composite and the ith component of the total of n components. For the case under study, $n = 2$, and the property P is Young's modulus. We can write, in extended form,

$$E_c = E_f V_f + E_m V_m \tag{10.3}$$

where the subscripts f and m indicate the fiber and matrix, respectively.

The elastic properties of such unidirectional composites in the transverse direction can be represented by a system of "action in series" [Figure 10.5(b)]. Upon loading in a direction transverse to the fibers, then, we have equal stress in the components. This model is equivalent to Reuss' classic treatment. We may write

$$\frac{1}{P_c} = \sum_{i=1}^{n} \frac{V_i}{P_i} \tag{10.4}$$

Once again, for the case of $n = 2$, and taking the property P to be Young's modulus, we obtain, for the composite,

$$\frac{1}{E_c} = \frac{V_f}{E_f} + \frac{V_m}{E_m} \tag{10.5}$$

The simple relations expressed Equation 10.4 and Equation 10.5 are commonly referred to as the "rule of mixtures." The reader is warned that this rule is nothing more than a first approximation; more elaborate models have been proposed. Following is a summary of various methods of obtaining composite properties:

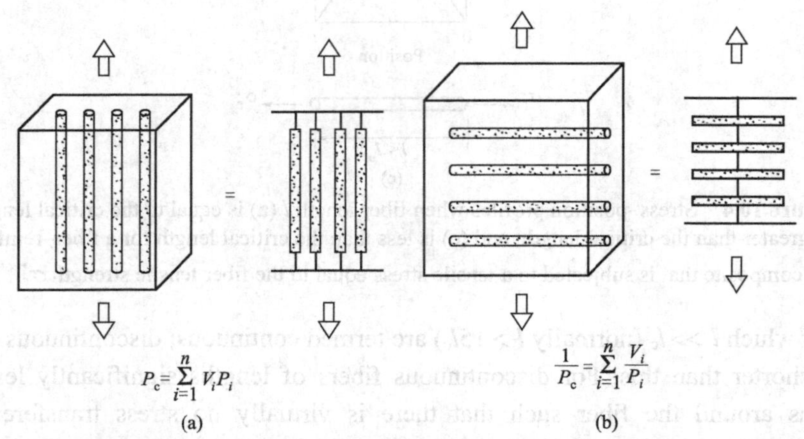

Figure 10.5 Simple composite models. (a) Longitudinal response (action in parallel). (b) Transverse response (action in series).

1. The mechanics-of-materials method. This deals with the specific geometric configuration of fibers in a matrix – for example, hexagonal, square, and rectangular – and we introduce large approximations in the resulting fields.

2. The self-consistent field method. This method introduces approximations in the geometry of the phases. We represent the phase geometry by a single fiber embedded in a material whose properties are equivalent to those of a matrix or an average of a composite. The resulting stress field is thus simplified.

3. The variational calculus method. This method focuses on the upper and lower limits of the properties of the composite and does not predict those properties directly. Only when the upper and the lower bounds coincide is a particular property determined. Frequently, however, the upper and lower bounds are well separated.

4. The numerical techniques method. Here we use series expansion, numerical analysis, and finite-element techniques.

10.2.2 Strength

Unlike elastic moduli, it is difficult to predict the strength of a composite by a simple rule-of-mixture type of relationship, because strength is a very structure-sensitive property. Specifically, for a composite containing continuous fibers and that is unidirectionally aligned and loaded in the fiber direction, the stress in the composite is written as

$$\sigma_c = \sigma_f V_f + \sigma_m V_m \tag{10.6}$$

where σ is the axial stress, V is the volume fraction, and the subscripts c, f, and m refer to the composite, fiber, and matrix, respectively. The reason that the rule of mixture does not work for properties such as strength, compared to its reasonable application in predicting properties such as Young's modulus in the longitudinal direction, is the following: The elastic modulus is a relatively structure-insensitive property, so, the response to an applied stress in the composite state is nothing but the volume- weighted average of the individual responses of the isolated components. Strength, on the contrary, is an extremely structure-sensitive property. Thus, synergism can occur in the composite state. Let us now consider the factors that may influence, in one way or the other, composite properties. First, the matrix or fiber structure may be altered during fabrication; and second, composite materials generally consist of two components whose thermomechanical properties are quite different. Hence, these materials suffer residual stresses and/or alterations in structure due to the internal stresses. The differential contraction that occurs when the material is cooled from the fabrication temperature to ambient temperature can lead to rather large thermal stresses, which, in turn, lead a soft metal matrix to undergo extensive plastic deformation. The deformation mode may also be influenced by rheological interaction between the components. The plastic constraint on the matrix due to the large difference in the Poisson's ratio of the matrix compared with that of the fiber, especially in the stage wherein the fiber deforms elastically while the matrix deforms plastically, can alter the stress state in the composite. Thus, the alteration in the microstructure of one or both of the components or the interaction between the components during straining can give rise to synergism in the strength properties of the composite. In view of this, the rule of mixture would be, in the best of the circumstances, a lower bound on the maximum stress of a composite.

Having made these observations about the applicability of the rule of mixture to the strength properties, we will still find it instructive to consider this lower bound on the mechanical behavior of the composite. We ignore any negative deviations from the rule of mixtures due to any misalignment of the fibers or due to the formation of a reaction product between fiber and matrix. Also, we assume that the components do not interact during straining and that these properties in the composite state are the same as those in the isolated state.

Then, for a series of composites with different fiber volume fractions, σ_c would be linearly dependent on V_f. Since $V_f + V_m = 1$, we can rewrite Equation 10.6 as

$$\sigma_c = \sigma_f V_f + \sigma_m(1 - V_f) \tag{10.7}$$

We can put certain restrictions on V_f in order to have real reinforcement. For this, a composite must have a certain minimum-fiber (continuous) volume fraction, V_{min}. Assuming that the fibers are identical and uniform (that is, all of them have the same ultimate tensile strength), the ultimate strength of the composite will be attained, ideally, at a strain equal to the strain corresponding to the ultimate stress of the fiber. Then, we have

$$\sigma_{cu} = \sigma_{fu} V_f + \sigma'_m (1 - V_f),\ V_f \geqslant V_{min} \tag{10.8}$$

where σ_{fu} is the ultimate tensile of stress of the fiber in the composite and σ'_m is the matrix stress at the strain corresponding to the fiber's ultimate tensile stress. Note that σ'_m is to be determined from the stress-strain curve of the matrix alone; that is, it is the matrix flow stress at a strain in the matrix equal to the breaking strain of the fiber. As already indicated, we are assuming that matrix stress-strain behavior in the composite is the same as in isolation. At low volume fractions, if a work-hardened matrix can counterbalance the loss of load-carrying capacity as a result of fiber breakage, the matrix will control the strength of the composite. Assuming that all the fibers break at the same time, in order to have a real reinforcement effect, one must satisfy the relation

$$\sigma_{cu} = \sigma_{fu} V_f + \sigma'_m (1 - V_f) \geqslant \sigma_{mu}(1 - V_f) \tag{10.9}$$

where σ_{mu} is the ultimate tensile stress of the matrix. The equality in this expression serves to define the minimum fiber volume fraction, V_{min}, that must be surpassed in order to have real reinforcement. In that case,

$$V_{min} = \frac{\sigma_{mu} - \sigma'_m}{\sigma_{fu} + \sigma_{mu} - \sigma'_m} \tag{10.10}$$

The value of V_{min} increases with decreasing fiber strength or increasing matrix strength.

In case we require that the composite strength should surpass the matrix ultimate stress, we can define a critical fiber volume fraction, V_{crit}, that must be exceeded. V_{crit} is given by the equation

$$\sigma_{cu} = \sigma_{fu} V_f + \sigma'_m (1 - V_f) \geqslant \sigma_{mu}$$

In this case,

$$V_{crit} = \frac{\sigma_{mu} - \sigma'_m}{\sigma_{fu} - \sigma'_m} \tag{10.11}$$

V_{crit} increases with increasing degree of matrix work-hardening ($\sigma_{mu} - \sigma'_m$). Figure 10.6 shows graphically the determination of V_{min} and V_{crit}. One notes that V_{crit} will always be greater than V_{min}.

In general, by incorporating fibers, we can increase the strength of the composite in the longitudinal direction. The strengthening effect in the transverse direction is not significant. Particle reinforcement can result in a more isotropic strengthening, provided that we have a uniform distribution of particles. Carbon, aramid, and glass fibers are used in epoxies to obtain high-strength composites. Such PMCs, however, have a maximum use temperature of about 150 ℃. Metal matrix composites, such as silicon carbide fiber in titanium, can take us to moderately high application temperatures. For applications requiring very high temperatures, we must resort to ceramic matrix

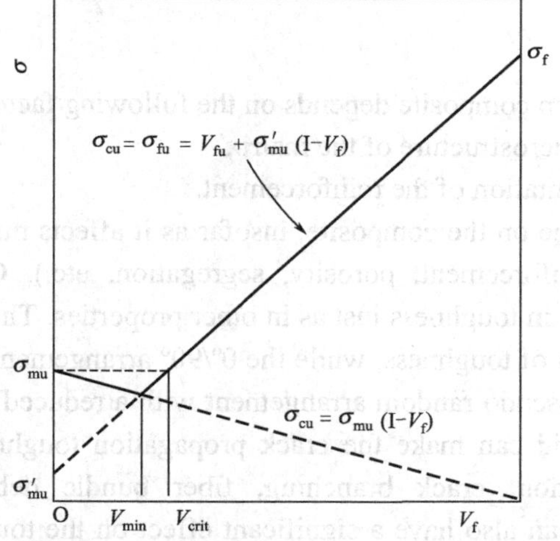

Figure 10.6 Determination of V_{min} and V_{crit}.

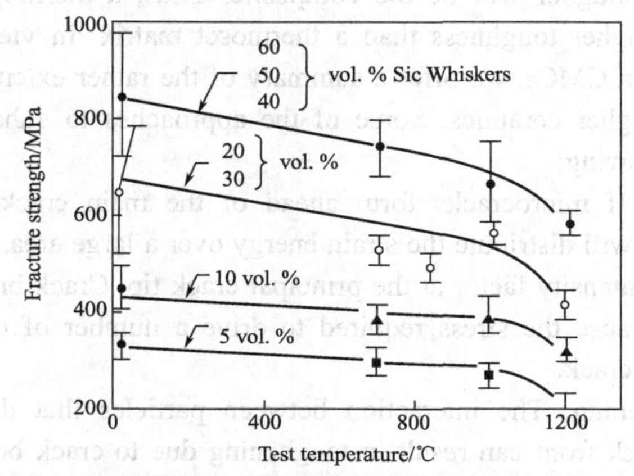

Figure 10.7 Increase in strength in silicon carbide whisker/alumina composites as a function of the whisker volume fraction and test temperature (After G. C. Wei and P. F. Becher, Am. Ceram. Soc. Bull., 64 (1985) 333) .

composites. Silicon carbide whisker reinforced alumina composites show a good combination of mechanical and thermal properties: substantially improved strength, fracture toughness, thermal shock resistance, and high-temperature creep resistance over that of monolithic alumina. Figure 10.7 gives an example of the improvement in strength in the silicon carbide whisker/alumina composites as a function of the whisker volume fraction and test temperature. Similar results have been obtained with silicon carbide whisker/reinforced silicon nitride composites. Finally, we should mention the strength of in-situ composites. In Figure 10.2(c), we showed the microstructure of an in situ composite. Such a composite is generally made by the unidirectional withdrawal of heat during the solidification of a eutectic alloy. This controlled solidification allows for one phase to appear in an aligned fibrous form in a matrix of the other phase. The strength σ of such an insitu metal matrix composites made by directional solidification of eutectic alloys is given by a relationship similar to the Hall-Petch relationship used for grain-boundary strengthening:

$$\sigma = \sigma_0 + k\lambda^{-1/2}.$$

Here σ_0 is a friction stress term, k is a material constant, and λ is the interfiber spacing between rods, or lamellae. It turns out that one can vary λ rather easily by controlling the solidification rate R, because $\lambda^2 R$ equals a constant. Thus, one can easily control the strength of these in-situ composites.

10.3 Toughness

The toughness of a given composite depends on the following factors:

(1) Composition and microstructure of the matrix.

(2) Type, size, and orientation of the reinforcement.

(3) Any processing done on the composite, insofar as it affects microstructural variables (e.g., the distribution of the reinforcement, porosity, segregation, etc.). Continuous fiber reinforced composites show anisotropy in toughness just as in other properties. The 0° and 90° arrangements of fibers result in two extremes of toughness, while the 0°/90° arrangement (i.e., alternating laminae of 0° and 90°) gives a sort of pseudo random arrangement with a reduced degree of anisotropy. Using fibers in the form of a braid can make the crack propagation toughness increase greatly due to extensive matrix deformation, crack branching, fiber bundle debonding, and pullout. The composition of the matrix can also have a significant effect on the toughness of a composite: The tougher the matrix, the tougher will be the composite. Thus, a thermoplastic matrix would be expected to provide a higher toughness than a thermoset matrix. In view of the importance of toughness enhancement in CMCs, we offer a summary of the rather extensive effort that has been expended in making tougher ceramics. Some of the approaches to enhancing the toughness of ceramics include the following:

(1) Microcracking. If microcracks form ahead of the main crack, they can cause crack branching, which in turn will distribute the strain energy over a large area. Such microcracking can thus decrease the stress intensity factor at the principal crack tip. Crack branching can also lead to enhanced toughness, because the stress required to drive a number of cracks is more than that required to drive a single crack.

(2) Particle Toughening. The interaction between particles that do not undergo a phase transformation and a crack front can result in toughening due to crack bowing between particles, crack deflection at the particle, and crack bridging by ductile particles. Incremental increases in toughness can also result from an appropriate thermal mismatch between particles and the matrix. Taya et al. examined the effect of thermal residual stress in a TiB_2 particle reinforced silicon carbide matrix composite. They attributed the increased crack growth resistance in the composite vis-à-vis the unreinforced SiC to the existence of compressive residual stress in the SiC matrix in the presence of TiB_2 particles.

(3) Transformation toughening. This involves a phase transformation of the second-phase particles at the crack tip with a shear and a dilational component, thus reducing the tensile stress concentration at the tip. In particulate composites, such as alumina containing partially stabilized zirconia, the change in volume associated with the phase transformation in zirconia particles is exploited to obtain enhanced toughness. In a partially stabilized zirconia (e.g., $ZrO_2 + Y_2O_3$), the stress field at the crack tip can cause a stress-induced martensitic transformation in ZrO_2 from a tetragonal phase (t) to a monoclinic one (m); that is,

$$ZrO_2(t) \rightarrow ZrO_2(m).$$

This transformation causes an expansion in volume (by approximately 4%) and a shear (0.16). The transformation in a particle at the crack tip results in stresses that tend to close the crack, and thus a portion of the energy that would go to fracture is spent in the stress-induced transformation. Also, the dilation in the transformed zone around a crack is opposed by the surrounding untransformed material, leading to compressive stresses that tend to close the crack. This results in increased toughness. Transformation in the wake of a crack can result in a closure force that tends to resist the crack opening displacement. Crack deflection at zirconia particles can also contribute to toughness.

(4) Fiber or whisker reinforcement. Toughening by long fibers or whiskers can bring into play a series of energy-absorbing mechanisms in the fracture process of CMCs and thus allow these materials to tolerate damage.

It appears that the effectiveness of various toughening mechanisms for structural ceramics decreases in the following order: continuous fiber reinforcement; transformation toughening; whiskers, platelets, and particles; microcracking. Many researchers have shown that if we add continuous C or SiC fibers to a glass or ceramic matrix, we can obtain a stress--strain curve of the type shown in Figure 10.10. This curve has the following salient features:

(1) Damage-tolerant behavior in a composite consisting of two brittle components.

(2) Initial elastic behavior.

(3) At a stress σ_0, the brittle matrix cracks.

(4) The crack bypasses the fibers and leaves them bridging the crack.

(5) Under continued loading, we have regularly spaced cracks in the matrix, bridged by the fibers.

(6) Noncatastrophic failure occurs. Fiber pullout occurs after the peak load, followed by failure of the composite when the fibers fail.

The final failure of the composite does not occur catastrophically with the passage of a single crack; that is, self-similar crack propagation does not occur. Thus, it is difficult to define an unambiguous fracture toughness value, such as a value for K_{Ic}. Figure 10.8 shows the stress vs. displacement curves for mullite fiber (Nextel 550)/mullite matrix. The uncoated one refers to the mullite/mullite composite with no interfacial coating. This composite shows a catastrophic failure. The composite containing a double interfacial coating of SiC and BN shows a noncatastrophic behavior because of the energy expending mechanisms such as interfacial debonding, fiber pullout, etc. come into play during the fracture process. The interfacial

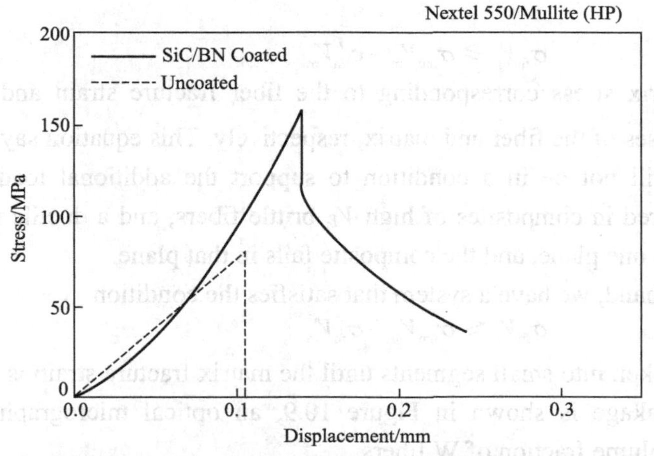

Figure 10.8 Stress vs. displacement curves for mullite fiber (Nextel 550)/mullite matrix in three-point bending. The uncoated one refers to the mullite/mullite composite with no interfacial coating, which shows a catastrophic failure. The composite with a double interfacial coating of SiC and BN shows a noncatastrophic [Adapted from K. K. Chawla, Z. R. Xu, and J.-S. Ha, J.Eur. Ceram. Soc., 16 (1996) 293] .

coatings provide for easy crack deflection, interfacial debonding, and fiber pullout.

It has been amply demonstrated that incorporation of continuous fibers such as carbon, alumina, silicon carbide, and mullite fibers in brittle matrix materials (e.g., cement, glass, and glass—ceramic matrix) can result in toughening. Not all of these failure mechanisms need operate simultaneously in a given fiber—matrix system, and often, in many composite systems, only one or two of the mechanisms will dominate the total fracture toughness.

10.4 Fracture in Composites

Fracture is a complex subject, even in monolithic materials. Undoubtedly, it is even more complex in composite materials. A great variety of deformation modes can lead to failure in a composite. The operative failure mode will depend, among other things, on loading conditions and the particular composite system. The microstructure has a very important role in the mechanics of rupture of a composite. For example, the fiber diameter, its volume fraction and alignment, damage due to thermal stresses that may develop during fabrication or service—all these factors can contribute to, and directly influence, crack initiation and propagation. A multiplicity of failure modes can exist in a composite under different loading conditions.

10.4.1 Single and Multiple Fracture

In general, the two components of a composite will have different values of strain to fracture. When the component that has the smaller breaking strain fractures, the load carried by this component is thrown onto the other one. If the latter component, which has a higher strain to fracture, can bear the additional load, the composite will show multiple fracture of the brittle component (the one with smaller fracture strain); eventually, a particular transverse section of composite becomes so weak, that the composite is unable to carry the load any further, and it fails. Let us consider the case of a fiber reinforced composite in which the fiber fracture strain is less than that of the matrix. Then the composite will show a single fracture when

$$\sigma_{fu}V_f \geqslant \sigma_{mu}V_m - \sigma'_m V_m \tag{10.12}$$

where σ'_m is thematrix stress corresponding to the fiber fracture strain and σ_{fu} and σ_{mu} are the ultimate tensile stresses of the fiber and matrix, respectively. This equation says that when the fibers break, the matrix will not be in a condition to support the additional load, a condition that is commonly encountered in composites of high V_f, brittle fibers, and a ductile matrix. All the fibers break in more or less one plane, and the composite fails in that plane.

If, on the other hand, we have a system that satisfies the condition

$$\sigma_{fu}V_f \leqslant \sigma_{mu}V_m - \sigma'_m V_m \tag{10.13}$$

the fibers will be broken into small segments until the matrix fracture strain is reached. An example of this type of breakage is shown in Figure 10.9, an optical micrograph of a Fe-Cu matrix containing a small volume fraction of W fibers.

In case the fibers have a fracture strain greater than that of the matrix (an epoxy resin reinforced with metallic wires), we would have a multiplicity of fractures in the matrix, and the condition for this may be written as

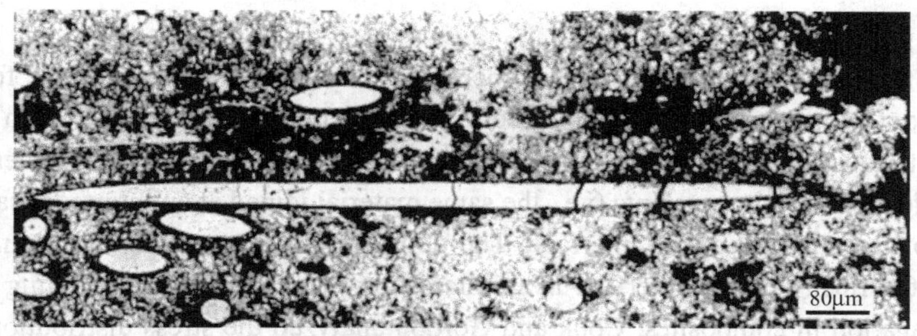

Figure 10.9 Optical micrograph of multiple fracture of tungsten fibers in an Fe-Cu matrix.

$$\sigma_{fu}V_f \geqslant \sigma_{mu}V_m - \sigma'_m V_f \tag{10.14}$$

where $\sigma'_f \sigma$ is now the fiber stress corresponding to the matrix fracture strain.

10.4.2 Failure Modes in Composites

Two failure modes are commonly encountered in composites:

(1) The fibers break in one plane, and, the soft matrix being unable to carry the load, the composite failure will occur in the plane of fiber fracture. This mode is more likely to be observed in composites that contain relatively high fiber volume fractions and fibers that are strong and brittle. The latter condition implies that the fibers do not show a distribution of strength with a large variance, but show a strength behavior that can be characterized by the Dirac delta function.

(2) When the adhesion between fibers and matrix is not sufficiently strong, the fibers may be pulled out of the matrix before failure of the composite. This fiber pullout results in the fiber failure surface being nonplanar.

More commonly, a mixture of these two modes is found: fiber fracture together with fiber pullout. Fibers invariably have defects distributed along their lengths and thus can break in regions above or below the crack tip. This leads to separation between the fiber and the matrix and, consequently, to fiber pullout with the crack opening up. Examples are shown in Figure 10.10.

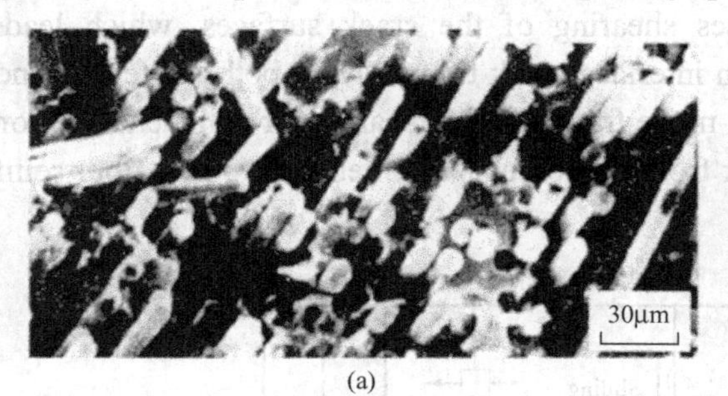

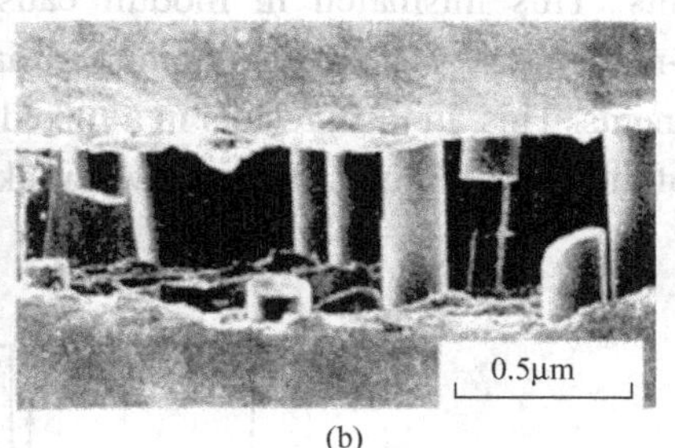

(a) (b)

Figure 10.10 Scanning electron micrographs of fracture in composites, showing the fiber pullout phenomenon. (a) Carbon fiber polyester. (b) Boron fiber aluminum 6061.

One of the attractive characteristics of composites is the possibility of obtaining an improved fracture toughness behavior together with high strength. Fracture toughness can be defined loosely as resistance to crack propagation. In a fibrous composite containing a crack transverse to the fibers, the crack propagation resistance can be increased by doing additional work by means of any or all of the following:

(1) Plastic deformation of the matrix.

(2) The presence of weak interfaces, fiber/matrix separation, and deflection of the crack.

(3) Fiber pullout.

It would appear that debonding of the fiber/matrix interface is a prerequisite for phenomena such as crack deflection, crack bridging by fibers, and fiber pullout. It is of interest to develop some criteria for interfacial debonding and crack deflection. Crack deflection at an interface between materials of identical elastic constants (i.e., the same material joined at an interface) can be analyzed on the basis of the strength of the interface. The deflection of the crack along an interface or the separation of the fiber—matrix interface is an interesting mechanism of augmenting the resistance to crack propagation in composites. Cook and Gordon analyzed the stress distribution in front of a crack tip and concluded that the maximum transverse tensile stress σ_{11} is about one-fifth of the maximum longitudinal tensile stress σ_{22}. They suggested, therefore, that when the ratio σ_{22}/σ_{11} is greater than 5, the fiber/matrix interface in front of the crack tip will fail under the influence of the transverse tensile stress, and the crack would be deflected 90° from its original direction. That way, the fiber/matrix interface would act as a crack arrester. This is shown schematically in Figure 10.11. The improvement in fracture toughness due to the presence of weak interfaces has been confirmed qualitatively.

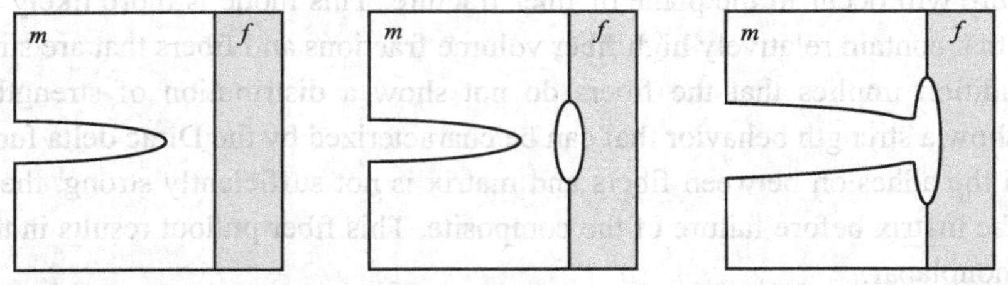

Figure 10.11 Fracture of weak interface in front of crack tip due to transverse tensile stress; *m* and *f* indicate the matrix and fiber, respectively [After J. Cook and J. E. Gordon, Proc. Roy. Soc. (London), A 228 (1964) 508]

Another treatment of this subject is based on a consideration of the fracture energy of the constituents. Two materials that meet at an interface are more than likely to have different elastic constants. This mismatch in moduli causes shearing of the crack surfaces, which leads to a mixed-mode stress state in the vicinity of an interface crack tip involving both the tensile and shear components. This, in turn, results in a mixed-mode fracture, which can occur at the crack tip or in the wake of the crack. Figure 10.12 shows crack front and crack wake debonding in a fiber reinforced

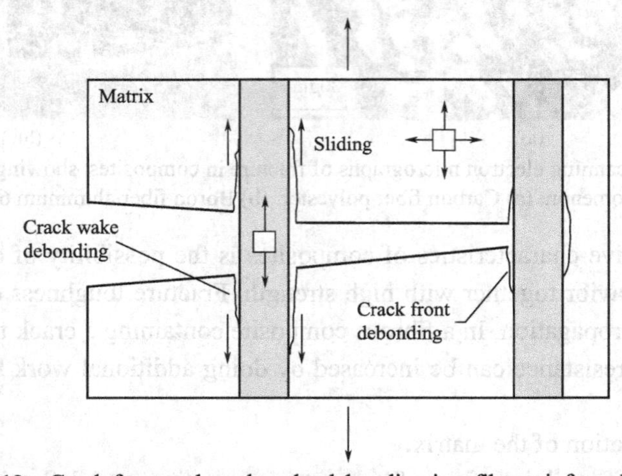

Figure 10.12 Crack front and crack wake debonding in a fiber reinforced composite.

composite. Because of the mixed-mode fracture, a single-parameter description by the critical stress intensity factor K_{Ic} will not do; instead, one needs a more complex formalism of fracture mechanics to describe the situation. In this case, the parameter K becomes scale sensitive, but the critical strain energy release rate G_{Ic} is not a scale-sensitive parameter. G is a function of the phase angle ψ, which, in turn, is a function of the normal and shear loading. For the opening mode, or mode Ⅰ, $\psi = 0°$, while for mode Ⅱ, $\psi = 90°$. One needs to specify both G and ψ to analyze the debonding at the interface. Without going into the details, we present here the final results of such an analysis, in the form of a plot of G_i/G_f vs. α, where G_i is the mixed-mode interfacial fracture energy of the interface, G_f is the mode- Ⅰ fracture energy of the fiber, and α is a measure of the elastic mismatch between the matrix and the reinforcement, defined as

$$\alpha = \frac{\overline{E}_1 - \overline{E}_2}{\overline{E}_1 + \overline{E}_2} \tag{10.15}$$

where

$$\overline{E} = \frac{E}{1-\upsilon^2} \tag{10.16}$$

The plot in Figure 10.13 shows the conditions under which the crack will deflect along the interface or propagate through the interface into the fiber. For all values of G_i/G_f below the cross-hatched boundary, interface debonding is predicted. For the special case of zero elastic mismatch (i.e., for $\alpha = 0$), the fiber/matrix interface will debond for G_i/G_f less than about 0.25. Conversely, for G_i/G_f greater than 0.25, the crack will propagate across the fiber. In general, for elastic mismatch, with α greater than zero, the minimum interfacial toughness required for interface debonding increases (i.e., high-modulus fibers tend to favor debonding). One shortcoming of this analysis is that it treats the fiber and matrix as isotropic materials; this is not always true, especially for carbon fiber.

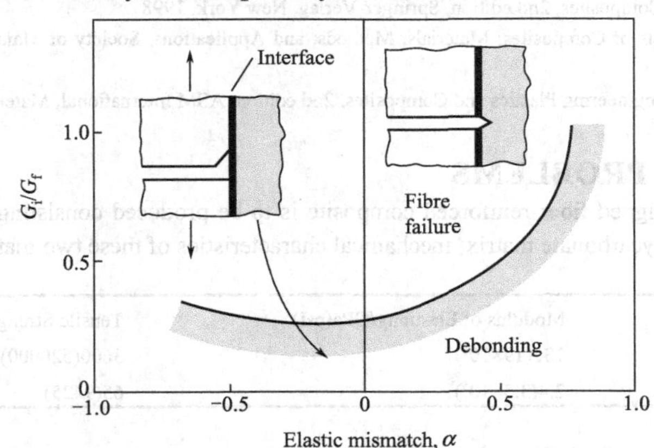

Figure 10.13 The ratio of the interface fracture toughness to that of fiber, G_i/G_f, vs. the elastic mismatch α. Interfacial debonding occurs under the curve, while for conditions above the curve, the crack propagates through the interface.

Gupta et al. derived strength and energy criteria for crack deflection at a fiber/matrix interface for several composite systems, taking due account of the anisotropic nature of the fiber. They used an experimental technique-spallation by means of a laser Doppler displacement interferometer-to measure the tensile strength of a planar interface. Through this technique, these researchers have tabulated the required values of the interface strength and fracture toughness for delamination in a number of ceramic, metal, intermetallic, and polymer matrix composites.

IMPORTANT TERMS AND CONCEPTS

Carbon–carbon composite	Ceramic-matrix composite	Whisker
Concrete	Dispersed phase	Dispersion-strengthened composite
Fiber	Fiber-reinforced composite	Hybrid composite
Laminar composite	Large-particle composite	Longitudinal direction
Matrix phase	Metal-matrix composite	Polymer-matrix composite
Transverse direction	Structural composite	Principle of combined action
Reinforced concrete	Specific modulus	Specific strength

REFERENCES

1. K. K. Chawla. Composite Materials: Science and Engineering, 2nd ed. New York: Springer, 1998
2. K. K. Chawla. Ceramic Matrix Composites, 2nd ed. Boston: Kluwer Academic, 1993
3. K. K. Chawla. Fibrous Materials. Cambridge, U.K.: Cambridge University Press, 1998
4. N. Chawla and K. K. Chawla, Metal Matrix Composites. New York: Springer, 2006
5. T.W. Clyne and P.Withers. Metal Matrix Composites. Cambridge, U.K.: Cambridge University Press, 1994
6. L. N. Phillips, ed. Design with Advanced Composite Materials. London: The Design Council, 1989
7. S. Suresh, A. Needleman, and A. M. Mortensen, eds. Fundamentals of Metal Matrix Composites. Stoneham, MA: Butterworth—Heinemann, 1993
8. Agarwal, B. D. and L. J. Broutman, Analysis and Performance of Fiber Composites, 2nd edition, Wiley, New York, 1990
9. Ashbee, K. H., Fundamental Principles of Fiber Reinforced Composites, 2nd edition, Technomic Publishing Company, Lancaster, PA, 1993
10. Chawla, K. K., Composite Materials Science and Engineering, 2nd edition, Springer-Verlag, New York, 1998
11. Chou, T. W., R. L. McCullough, and R. B. Pipes, "Composites," Scientific American, Vol. 255, No. 4, October 1986, pp. 192–203
12. Hollaway, L. (Editor), Handbook of Polymer Composites for Engineers, Technomic Publishing Company, Lancaster, PA, 1994
13. Hull, D. and T.W. Clyne, An Introduction to Composite Materials, 2nd edition, Cambridge University Press, New York, 1996
14. Mallick, P. K., Fiber-Reinforced Composites, Materials, Manufacturing, and Design, 2nd edition, Marcel Dekker, New York, 1993
15. Peters, S.T., Handbook of Composites, 2nd edition, Springer-Verlag, New York, 1998
16. Strong, A. B., Fundamentals of Composites: Materials, Methods, and Applications, Society of Manufacturing Engineers, Dearborn, MI, 1989
17. Woishnis,W. A. (Editor), Engineering Plastics and Composites, 2nd edition, ASM International, Materials Park, OH, 1993

QUESTIONS AND PROBLEMS

10.1 A continuous and aligned fiber-reinforced composite is to be produced consisting of 45 vol% aramid fibers and 55 vol% of a polycarbonate matrix; mechanical characteristics of these two materials are as follows:

	Modulus of Elasticity/GPa(psi)	Tensile Strength/MPa(psi)
Aramid fiber	131(19×10^6)	3600(520000)
Polycarbonate	2.4(3.5×10^5)	65(9425)

Also, the stress on the polycarbonate matrix when the aramid fibers fail is 35MPa (5075 psi).
For this composite, compute:
(a) the longitudinal tensile strength,
(b) the longitudinal modulus of elasticity.

10.2 The mechanical properties of cobalt may be improved by incorporating fine particles of tungsten carbide (WC). Given that the moduli of elasticity of these materials are, respectively, 200GPa (30×10^6psi) and 700 GPa(102×10^6psi), plot modulus of elasticity versus the volume percent of WC in Co from 0 to 100 vol%, using both upper- and lowerbound expressions.

10.3 Estimate the maximum and minimum thermal conductivity values for a cermet that contains 90 vol% titanium carbide (TiC) particles in a nickel matrix. Assume thermal conductivities of 27 and 67 W/m-K for TiC and Ni, respectively.

10.4 Cite one similarity and two differences between precipitation hardening and dispersion strengthening.

10.5 (a) What is the distinction between cement and concrete?

(b) Cite three important limitations that restrict the use of concrete as a structural material.

(c) Briefly explain three techniques that are utilized to strengthen concrete by reinforcement.

10.6 For a continuous and oriented fiberreinforced composite, the moduli of elasticity in the longitudinal and transverse directions are 33.1 and 3.66 GPa (4.8×10^6 and 5.3×10^6 psi), respectively. If the volume fraction of fibers is 0.30, determine the moduli of elasticity of fiber and matrix phases.

Chapter 10

IMPORTANT TERMS AND CONCEPTS

Principle Of Combined Action	复合原则	Whisker	晶须
Carbon–Carbon Composite	碳碳复合材料	Dispersion-strengthened Composite	弥散强化复合材料
Laminar Composite	层状复合材料	Hybrid Composite/ Miscellaneous Composite	混杂复合材料
Matrix Phase	基体相	Dispersed Phase	弥散相
Transverse Direction	横向	Short Fiber Reinforcement	短纤维增强
Lengthwise Direction/ Longitudinal Direction	纵向	Discontinuous Fiber Reinforcement	非连续纤维增强
Ceramic-matrix Composite, CMC	陶瓷基复合材料	Continuous Fiber Reinforcement	连续纤维增强
Polymer-matrix Composite, PMC	聚合物基复合材料	Aramid Fiber	芳族聚酰胺纤维
Metal-matrix Composite, MMC	金属基复合材料	Polyester	聚合酯
Fiber-reinforced Composite	纤维增强复合材料	Delamination/ Spallation	剥落/分裂
Particle-reinforcement Composite	颗粒增强复合材料	Debonder	剥离器
Laminate-reinforcement Composite	编织增强复合材料	Crack Front Debonding	裂纹前端剥离
Pullout	拔拉	Crack Wake Debonding	裂纹尾部剥离